Rethinking Generosity

CONTESTATIONS

A series edited by

WILLIAM E. CONNOLLY

A complete list of titles in the series appears at the end of the book.

Rethinking Generosity

Critical Theory and the Politics of Caritas

Romand Coles

Cornell University Press

Ithaca and London

First published 1997 by Cornell Univesity Press.

Printed in the United States of America.

Cornell University Press strives to utilize environmentally responsible suppliers and materials to the fullest extent possible in the publishing of its books. Such materials include vegetable-based, low-VOC inks and acid-free papers that are also either recycled, totally chlorine-free, or partly composed of nonwood fibers.

Library of Congress Cataloging-in-Publication Data

Coles, Romand, b. 1959
    Rethinking generosity : Critical theory and politics of caritas /
Romand Coles.
        p.   cm.
    Includes bibliographical references and index.
    ISBN 0-8014-3341-X (cloth: alk. paper) ISBN 0-8014-8487-1 (pbk.: alk. paper)
    1. Generosity.   2. Critical theory.   I. Title.   II. Series.
BJ1533.G4C65   1997
177'.7—DC21   97-18760

Cloth printing   10 9 8 7 6 5 4 3 2 1
Paperback printing   10 9 8 7 6 5 4 3 2 1

Contents

Preface

These are not generous times. Yet as I read the human condition, receptive generosity is integral to—even definitive of—our thriving. In its absence human being wanes.

This waning is not simply a consequence of political economies, cultures, and ethics that cultivate narrow self-interest. It also happens when efforts to practice generosity become blind to the way in which the possibility of generosity is entwined with a difficult receptivity toward others who are different. When generosity becomes separated from receptivity, it tends toward imperialism and theft. We should not be too surprised, then, when such generosity forms alliances with more explicitly stingy practices. To articulate a vision of receptive generosity is to resist both possessive individualism and ideologies of monological generosity, and to seek to contribute to the possibility of something better.

In the Introduction, I elaborate the ethical and political stakes involved in questions concerning receptive generosity. I then provide a philosophical overview of some of the problems of receptivity which arise with secularization and which are confronted in various early modern thinkers. Any effort to explore receptive generosity must struggle with the secular response to these problems that takes shape in the paradigm of sovereign subjectivity which has pervasively influenced ethical reflection in our age. Thus Kant, who formulated the most powerful articulation of this project, is a key interlocutor in this book. I read Kant's prefatory remarks to the *Critique of Pure Reason* as a drama in which Kant writes himself as the epic hero of the secular, brilliantly

defining the problems he faces and the direction and limits of his philosophy of sovereign subjectivity. I close the introduction with a reading of Nietzsche's *Thus Spoke Zarathustra*, in an effort to juxtapose to
Kant's story a tale in which sovereign subjectivity undergoes a tragic
unraveling in ways which suggest and inspire my efforts toward a philosophy and politics of receptive generosity.

Chapter 1 focuses on Kant's paradigmatic attempts to formulate the
theme of giving within the context of sovereign subjectivity. He unwittingly reveals the untenability of much of his vision. I conclude this
chapter by drawing on aspects of the *Critique of Judgment* which suggest a notion of giving entwined with receiving, with which I recon-
ceive some of Kant's reflections on the public sphere. Kant's more
buoyant solicitation of post-secular *caritas* informs my reading of
Adorno, but also remains in tension with Adorno's more melancholy
accents.

Chapter 2 focuses on Adorno, who arguably offers the most powerful critique of the problems associated with subjective sovereignty. Less
recognized is that Adorno's negative dialectics offers perhaps the most
profound reformulation of generosity as receptive generosity. Freeing
Adorno from numerous contemporary misreadings, I illuminate receptivity, generosity, and the problems of dialogue as central animating
themes in his work. His writing offers an ethical constellation of diverse solicitations that are agonistically juxtaposed in order to provoke
and nurture receptive generosity.

In Chapter 3, I engage the work of Habermas, who is not only diametrically opposed to such a reading of Adorno, but moreover understands (an understanding echoed by many) his own position as the best
effort to elaborate the most generous suggestions in Kant's work in a
manner that escapes the weaknesses and dangers of Kantian subjectivity. Following an overview of Habermas's theory of communicative
action and discourse ethics, I develop a critical reading of Habermas, in
the light of my account of Adorno, to identify and move beyond many
of the weaknesses in communicative ethics.

In the Conclusion, I shift to a more explicitly political terrain. I begin with a discussion of generosity and receptivity in Marx, who sought
to bring these virtues into history. Abandoning the agon-free ethics and
politics of collective subjectivity that frame some of Marx's vision, I
suggest that a transfigured politics of receptive generosity must work

and develop itself throught a politics of partly agonistic coalitions among diverse groups. My reading of Adorno provides a promising ethical vision for addressing some of the important difficulties and motivational needs of people doing coalition politics, as well as for reawakening a more vital civil society. By significantly reformulating some of Habermas's political, economic, and social insights in the light of the ethic I read in Adorno, important possibilities emerge that Adorno did not explore.

In the sections on Kant's first and second *Critiques*, the recounting of the Habermasian paradigm, as well as in my discussions of Marx, and Laclau and Mouffe, I have tried to give as accurate a reading of the texts as I can. The sections on Nietzsche, Adorno, Reagon, and Kant's Third Critique are the manifestation of processes of reciprocal transfiguration between myself and the texts, and are driven by the questions at hand. Though often difficult to separate, the obligation to accuracy has had to contest with the obligation to think *with* others in directions they richly suggest but do not fully pursue.

Here I acknowledge and thank a few of the many people who contributed to this project. Bill Connolly—far beyond any call of editorial duty—offered detailed comments on the entire manuscript, and energetically and insightfully discussed numerous issues at length. Roger Cooper, William Corlett, Kimberly Curtis, Frederick Dolan, Michael Gillespie, Bonnie Honig, Michael James, and Steven White, generously offered invaluable comments and criticisms on portions or all of this work. Traces of many other conversations are evident to me on nearly every page. This book would have been impossible without the continuing inspiration, support, and challenges of the graduate students in my seminars at Duke University during the past eight years. Roger Haydon offered very sharp editorial advice from which this book benefited significantly. John Ackerman offered numerous brilliant suggestions on my previous book with Cornell University Press, and I somehow forgot to acknowledge this publicly. I hope the regrettable delay of this expression of gratitude somehow adds to its depth. Charles Purrenhage's copyediting improved this text. Thanks to Johnny Goldfinger for his sharp eyes and gracious perseverance with the page proofs. Joy Pickett patiently and skillfully helped type the many drafts of this book.

Two people in my family died while I was writing this book, my grandmother, Josephine Iannacome, and my grandfather, Abraham De-Nearing. Two were born, my children, Marianna and Miles Coles-Curtis. I dedicate this book to the four of them.

ROMAND COLES

Durham, North Carolina

Abbreviations

In the Kant chapter, because I deal with single texts at length, I simply use page numbers in parentheses where the text is obvious. In the Adorno and Habermas chapters, and in the Conclusion, I use the following abbreviations. For texts less cited, I use traditional endnote format.

THEODOR ADORNO

AT   *Aesthetic Theory.* Trans. Christian Lenhardt. New York: Routledge and
        Kegan Paul, 1984.

DE   With Max Horkheimer. *Dialectic of Enlightenment.* Trans. John Cum-
        ming. New York: Seabury Press, 1972.

MM   *Minima Moralia: Reflections from Damaged Life.* Trans. E. F. N. Jephcott.
        London: Verso, 1974.

ND   *Negative Dialectics.* Trans. E. B. Ashton. New York: Continuum, 1983.

JÜRGEN HABERMAS

AS   *Autonomy and Solidarity: Interviews.* Ed. Peter Dews. London: Verso,
        1986.

"DE"   "Discourse Ethics," in *MCCA.*

FN   *Between Facts and Norms: Contributions to a Discourse Theory of Law and
        Democracy.* Trans. William Rehg. Cambridge: MIT Press, 1996.

JA   *Justification and Application: Remarks on Discourse Ethics.* Trans. Ciaran P.
        Cronin. Cambridge: MIT Press, 1993.

KHI   *Knowledge and Human Interests*. Trans. Jeremy Shapiro. Boston: Beacon Press, 1971.

MCCA   *Moral Consciousness and Communicative Action*. Trans. Christian Lenhardt and Sherry Weber Nicholsen. Cambridge: MIT Press, 1990.

PDM   *The Philosophical Discourse of Modernity: Twelve Lectures*. Trans. Frederick Lawrence. Cambridge: MIT Press, 1987.

PPP   *Philosophical-Political Profiles*. Trans. Frederick Lawrence. Cambridge: MIT Press, 1983.

PT   *Postmetaphysical Thinking*. Trans. William Mark Hohengarten. Cambridge: MIT Press, 1992.

TCA 1   *The Theory of Communicative Action*. Vol. 1: *Reason and the Rationalization of Society*. Trans. Thomas McCarthy. Boston: Beacon Press, 1984.

TCA 2   *The Theory of Communicative Action*. Vol. 2: *Lifeworld and System*. Trans. Thomas McCarthy. Boston: Beacon Press, 1987.

Rethinking Generosity

Questioning *Caritas*

This book explores possibilities for an ethic and a politics of receptive generosity. It arises from my sense that movements of receptive generosity to and fro among diverse human beings articulate much of what is most admirable about us. Yet it is also driven by a threatening shadow, the sense that generosity is perhaps our most difficult endeavor. Not only do we seem under many conditions to have weak, intermittent, and hypocritical attachments to this ethical call, but even when we do fix our sights on it, we seem prone to engender so little, perhaps nothing, maybe even theft or murder. Indeed, it is difficult to write of generosity today without conjuring up images of the terror wrought by a religion that at once placed the movement of *caritas* and *agape*, giving and love, at the foundation of being and swept across the Americas during the Conquest with a holocaust of "generosity."[1]

Hopes and Fears: *Caritas* beyond *Caritas*

It would be nice, wouldn't it, if we could easily separate generosity from holocaust and theft, if we could say that in this or that holocaust generosity was simply forgotten or insincerely evoked? But what if the matter is more complicated? What if there is something about generosity, or at least about its dominant modes of articulation, that leads it to forget itself: not in the more admirable and even necessary sense of the forgetting attributable to a consuming commitment that overshadows self-consciousness, but a forgetting that turns away from the spirit of the project entirely and tends rather in directions of pillage and death? This latter at any rate is my suspicion. And though I conceive of this as a hopeful book, it is a hope borne by a sense of danger. The reformulations of generosity that follow do not expel that danger; rather, they suggest a constellation of concerns, insights, and solicitations that might enhance our capacities to address it.

Dominant modes of imagining and understanding generosity in Western history—I have in mind here particularly many of those rooted in various forms of either Christianity or modern rational subjectivity—have been constituted on the idea of a self-identical ground from which is given all being, truth, moral value, and beauty. Perhaps this self-origination is sometimes understood better as an internally differentiated dynamic flow, a giving always moving beyond itself, and in this sense less stable than a self-identity. Yet even as in some sense dynamic, *even in its movement*, the fount of giving is self-identical insofar as this flow does not receive—and even precludes reception of—alterity. The archetypes of giving have been profoundly divorced from receiving. Many understandings do not admit the possibility that God or the transcendental subject can receive anything radically other.

Think, for example, of Augustine's formulation of God, a formulation that establishes the outline of faiths that have operated for centuries. God is the very movement of *caritas* and *agape*, and these qualities infuse the being of his gift: namely, all of creation as the temporal elaboration of his word. We, of his loving gift, have been given his Son, who through the Gospels exemplifies the incarnation of *caritas* and teaches us how to receive God's love and in turn proliferate giving. To receive and follow this path, to faithfully interpret God's signs and love one's neighbor, is to "be fruitful and multiply," as Augustine writes in one of his most provocative moments. It is thus through receiving and giving that we participate in the unfolding of his gift and being; whereas to reject the flow of *caritas* is to tend toward nothingness.[2]

From one perspective, the movement of *caritas* is receptive of, generous toward, and in fact proliferates the wild multiplicitous diversity of creation. This sensibility is not hard to find in Augustine's writings, and it resounds in the work of many contemporary Augustinians who accent this dimension of their faith.[3] Christian community is to embrace the uniqueness of all its members as particular mysterious expressions of God's gift: it is not a vision of homogeneity. Yet the problem is that *these forms* of Christian receiving, giving, and proliferation are based on an imagination profoundly blind to the possible being and value of radical alterity in people who live resolutely outside the Christian story. That which lies beyond or rejects the Christian tale is, in this beyondness or rejection, understood as tending toward nothingness and evil. This malignancy at the heart of the loving gift rears its head in *City of God*, when Augustine writes that to love my neighbor as myself is to try to give him or her what I myself most need and want, namely the will to love God. Hence, the generosity engendered by the Augustinian Christian narrative is one whose overarching aim is given entirely from within itself. The profoundest goals, meanings, and textures of the generous act do not emerge from a receptive encounter with the other that risks—through opening onto aspects of alien needs, perceptions, desires, pleasures, and understandings—transgressing the self-enclosures of one's own narratives in ways that might transfigure the movement of giving. Most fundamentally, such generosity is not wrought in the unpredictable interstice between Christian selves and non-Christian others; it is determined prior to and over against the encounter such that the other never appears in its otherness. The other is either nothingness and evil or a moving toward inclusion within the same.[4]

Such an orientation is destined to manifest itself repeatedly, as it has already for centuries, in ugly encounters: violent conflicts within the church and at its boundaries. Insofar as generosity does not understand itself to be deeply rooted in a receptive encounter with others, it will proliferate a blindness, theft, and imperialism despite its best efforts; it will ensure an oblivion that continually suppresses the question concerning how intelligent giving might happen, given the myriad specificities of the moment that calls for the gift. The most difficult and often the highest aspect of giving is receiving the other in agonistic dialogical engagements. Such engagement is not reducible to an a priori injunction to "let be." Rather, it is an effort to erode a priori closures so that the play of mutual transfigurations which are a condition of possibility for sense, intelligence, and well-being might thrive.

Christian narratives are by no means the only forms of self-defeating generosity based on an ungenerous ontology that too heavily accents identity. Dominant formulations of the modern rational subject, exemplified most powerfully by Kant, pursue a project whose efforts are animated by and in important ways remain within the limits of a secularized version of Christian (especially Ockhamist) narrative. Hence the rational subject is understood most fundamentally as that which auton- omously "gives the law to nature" and "gives itself the moral law." This rational, monological subjective giving that is said to be constitutive of experience, understanding, and a morality of respect for other such autonomous subjects faces problems analogous to those to which I have alluded above. Kantian sovereign subjects are to give themselves the moral law such that they give respect to each human being. On one level, Kant takes pains to warn against humiliating others, calls us to avoid imperialism in giving and to respect *others'* concepts of happiness in our beneficence, and he seeks to deflate the sense of pride that often accompanies giving.[5] Yet how likely is the Kantian self (which funda- mentally understands its moral reason in a manner that is not only devoid of but hostile to the reception of alterity) to position and articu- late itself with respect to others in a manner such that generosity is even conceivable?

It will perhaps be objected that my reading of "giving the law" in terms of questions of generosity is based on a misconstrual: the terms are the same, but the themes are very different insofar as "giving" in the transcendental narrative concerns the possibility of the experiential and moral worlds, not generosity in the first instance. Yet the connec- tions are far more than terminological. First, Kant's *Critique of Practical Reason* does attempt, through his understanding of "giving the law," to reinfuse moral being with a kind of generosity that had been radically problematized both by late-medieval understandings of omnipotence and by subsequent responses. Second, I shall show that our efforts to understand, articulate, and practice generosity are substantially influ- enced by our assumptions and basic narratives concerning how our being and that of the world are "given." In this light, Kant's transcen- dental narrative of the sovereign subject works to undermine more ad- mirable aspects of his thinking, and "respect" is entwined with a sys- tematic kind of oblivion, imperialism, and theft. As I read things, there are moments of strong complicity between the Kantian epistemological and moral foundational narratives and the frequent failures of modern liberal societies concerning otherness. The liberal violent conquest of "uncivilized peoples," the efforts to assimilate those who remained alive, the myriad efforts to constitute normalized are "respectable" ra- tional subjects, and the frequent fetishization of a moral righteousness and autonomy that is often relatively oblivious and inactive in its engagement with the world resonate with the deepest aspects of Kant's philosophy, despite so much in Kant that is better than this.

I do not wish simplistically to reduce Augustinian Christianity or modern Kantian subjectivity. I see things that are admirable and compelling in both narratives as well as in some of the practices that have been associated with them. But I also see a deleterious oblivion, cultivated and systematically concealed by foundational narratives hostile to receptivity. In the stereoscopic depth born at the intersection of these two lines of vision and a manifold world, in this commingling of attractions and repulsions, I find myself drawn to new horizons. These directions and my efforts to articulate them are significantly indebted to a terrain that is deeply permeated by Christianity and modern autonomous subjectivity. Yet they go beyond this terrain: partly as efforts to explore other regions, but perhaps most importantly as transfigurations of the terrain itself, particularly with respect to the entwinement of giving and being.

Hence, in the midst of my dissatisfactions, by employing the word *caritas*, I acknowledge a debt. But I wish to resist two misconstruals of this debt. The first suggests that the debt signifies the true impossibility and hopelessness of such a project. By these lights, *caritas* is inextricably a part of Christian narratives and their omnipotent charitable God, outside of which it is inconceivable. Even if such *caritas* were conceivable, some would argue, it would be utterly uncompelling without an omniscient rewarding and punishing God to command it. The second misconstrual implies that with my affirmation of indebtedness I thereby recognize the historical contingency of my claim, removing myself from the realm of ethical claims that may in any way exceed this facticity.

The first construal, in its crudest form, misses because, with a paradigmatic lack of receptivity, it assumes the very impossibility (of alternatives) that it would need to prove concretely in each encounter with other articulations and ways of being. Alternatives to the "original" narrative consist of either *wholly* other efforts to construe the practice of giving as central to human existence[6] or *partially* other efforts that bring aspects of Christianity into a constellation of insights that transfigures the meaning of those aspects, that draws on other narratives for support and power, and that might develop counterpractices through which these alternatives are given more texture. History is replete with such transfigurations. Their power and desirability vary widely, and each transfiguration calls for very specific engagement and judgment, as does the current effort. In this light, the voice of impossibility appears as an exemplification of the a priori taboo on receptivity.

Yet there are subtler forms of this basic claim which are more compelling. For example, Charles Taylor, in *Sources of the Self*, masterfully surveys modern versions of morality that reject any theological ground, and he makes a powerful case that they variously erode their own conditions of possibility, tacitly rely on theological positions they explicitly reject, are fraught with concealed contradictions, and so forth. Is it possible to draw a strong ethical sensibility in a world lacking even the whispers of God discerned by some of the romantic expressivists? My sense is that it is, and this book is largely an effort to do so. In the concluding section of the chapter on Adorno, I gather together themes in his work in an effort to respond directly to some of Taylor's challenges.

The second construal misses because it too assumes what it would need to prove, but cannot, given its own assumptions. From the idea that all human utterances and questions are profoundly spurred and colored by the contingencies of history—an idea I find persuasive—it is a big leap to claim that we can perceive, think, and write nothing that exceeds or opens beyond currently received contingencies and their history. In truth, the compelling arguments concerning historical finitude (which are prominent in the geography within which my ethical reflections move) indicate not the certainty of total historical determinacy and delimitations (from what perspective could such certainty be secured?) but, rather, an essential "undecidability" concerning the object and status of claims in this domain.[7] We are marked too deeply by historical finitude to claim smugly either "This is true" or "This has nothing to do with anything but the contingency that 'I just happen to believe.'" We open our senses, we think and speak, and we just don't know with certainty what it means or even what "meaning" means. We seem at once powerfully compelled by something like what Merleau-Ponty called a "perceptual faith" that we open (through finitude) onto a world that partially transcends the finitude of our contingent being as bodily and historical, while at the same time we repeatedly experience the limiting, radically blinding, and tragically damaging sway of this very finitude. These sensibilities don't often meet in a happy middle. They fight it out. Hence undecidability.

In this condition we can, indeed we must, consider ethical questions seriously. We continue to be ethically solicited; and we strive to articulate the opening of ethical possibility, and even an ethic, because nothing appears more compelling (in the light of myriad narratives, a few of which I discuss below) than that *we must*. And yet, in the face of undecidability, we cannot shake the ambiguity that shadows our efforts. Perhaps we should give up trying to shake it and instead allow it to inform our every move. Perhaps we should bear this shadow like a compulsion—like a man I once saw on the streets of Berkeley who seemed to be living with anticipatory concern for how his shadow would accompany each step, each gesture. What I mean is that we might articulate an ethical constellation that strives to recollect the questionability of its directions in its ongoing effort at ethical navigation. This would mean that in responding to the injunction *to give*, we would strive to be aware of the ambiguities and possible damage that will almost certainly accompany our every move, and we would do so in a manner that transfigures giving into a fundamentally interrogative task of receptivity. We ought to strive to articulate and practice ethics such that we repeatedly ask, What will happen to our questionability? Will it grow too dim with this declaration, this act? How can we speak and act in ways that keep this shadow alive? For if we lose *this*, we lose receptivity, generosity, ethics.

Secular Questions, Secular Strivings

If the word *caritas* acknowledges a historical debt, perhaps "post-secular" has a more anticipatory ring. The tension between these two words is meant to check the misconstruals that accompany each term in isolation. One hopes that the overly heady aura of "post-secular" will be subdued a bit in juxtaposition with "*caritas*." But what do I mean by "post-secular"? And by "secular"?

There are numerous debates concerning what is often called secularization, though they do not all employ the term.[8] John Milbank argues convincingly that late-medieval religious developments powerfully articulated in William of Ockham constituted a heretical turn in relation to the predominant late-antique and medieval forms of Christianity, which comprehended God's omnipotence to be essentially expressed in proliferating trinitarian relations of *caritas* and *agape*. With Ockham, a more singular sovereignty supplants the Trinity, and omnipotence is intensified such that it rips free of its essential inscription in the constellation with love and charity. God's omnipotence becomes radically singular and contingent: capable of changing the past, potentially capable of malicious deception. The world in turn sheds the veil of "gift" and becomes the stuff, place, and occasion of deception. As God and the world recede from the terrain of *caritas*, they elicit not receptivity but intensifying skepticism and fear.

The secular might be conceived in relation to the crisis provoked by this late-medieval synthesis in two senses. First, it is a response to questions prompted by the crisis concerning reception. Second, it attempts in diverse ways to place human beings (and their political and economic institutions) on the sovereign throne of the dying heretical God. Secular efforts adopt many of the contours of the very religion that provokes the crisis to which they are a response. Since Kant is the exemplary manifestation of the secular, insofar as he tries to articulate the sovereign human subject in a manner powerfully cognizant of the pitfalls associated both with previous such efforts and with previous rejections of the project of constructing human sovereignty, I shall focus extensively on his work. For the present, however, I wish to situate Kant's project with the briefest discussion of a few predecessors who help clarify some of the problems he faced that are of particular importance to my project.[9]

It is an Ockhamist world that Descartes sees as he glances around his study and out the window to the street below. Since God may be an evil deceiver and the world—as *res extensa*—is no longer directly accessible to our knowledge, Descartes seeks secure foundations within himself in the *cogito*. With this move, he takes a crucial step toward a subjectivity whose being is *most fundamentally* neither a gift from God nor essentially entwined in giving and receiving within God's creation; rather, it is *a gift the self gives itself* in the mere act of thinking. Without getting mired in the debates about the role of God in Descartes's thought, what is crucial here is that the "I think" must come *before* his positing of God which guarantees the possibility of harmony between thought and world. Even if the reception of God's gift in the form of a metaphysics of nondeception does become important, it is secondary to self-positing certainty in the first instance.

Yet Descartes's bridging of the gap between *res cogitans* and *res extensa* proved unsatisfactory to many who came after him. Berkeley's critique of the link between the belief in mind-independent "external bodies" and skepticism applies just as sharply to popular understandings of the Cartesian *res extensa* as it does to such "materialists" as Locke and Newton, with whom he was most immediately concerned. Indeed, Berkeley claims that "the arguments urged by skepticism in all ages depend on the supposition of external objects."[10] In terms of my narrative, when the being of the external world ceases to be fundamentally a gift from God that is receivable by humans through a metaphysics of analogy, resemblance, the Scholastic doctrine of the migrating properties of intentional species, and so forth, the abyss between mind and world appears unbridgeable. As Berkeley argues, those who retain a faith in real things "could not be certain they had any real knowledge at all. For how can it be known that the things which are perceived are conformable to those which are not perceived or exist outside the mind?"[11] The idea of secure possibilities of communication between spirit and external objects becomes unfathomable. Yet, instead of ending in skepticism, this very unfathomability prompts Berkeley to disregard the idea of external objects altogether and to identify the "real" with objects of perception in his famous "*esse* is *percipi*." To claim that "the existence of an idea [or thing] consists in being perceived" is to maintain that it is a passive dependent correlate of an active consciousness which gives it being, and thereby overcomes the problem of unreceivable otherness.

Perhaps too well. Inherent in this originary activity is a notion of *utterly arbitrary power* ("I can excite ideas in my mind at pleasure . . . vary . . . shift . . . as I think fit") that discovers its inability to account for a world of recalcitrance, continuity, and regularity ("I find the ideas actually perceived by sense have not a like dependence on my will").[12] In response, a supremely active perceiver is posited as he who produces the sensations we receive. Berkeley introduces the mind of some "External Spirit"—an all-wise, all-good, and all-powerful God who produces and actively sustains these ideas of sensation.[13] Natural law and mundane necessity become the effects of this benevolent will. With this move, though, Berkeley appears to return to the very being whose increasing absence is provocatively intertwined with his own skeptical reflections on disenchanted matter.

Berkeley's introduction of God has struck many commentators as problematic and arbitrary. As Buchdahl has noted, it "comes close to replacing the words 'physical external object' by 'God's power.'"[14] What is key, however, is that God is preferable to external objects be-

cause we can generate the former idea (and the ideas of "others" generally) with logical consistency "by means of our own soul."[15] That is to
say, the notion of God is generated through an active analogy with our
experience of our own active being, and requires less of a leap—or
none at all, for Berkeley. Thus the soul retains a dimension of epistemological priority. Though Berkeley concedes that God is the conscious power that grounds the perceptibility and being of the world, he
can still argue in the "Second Dialogue" that the being of God is arrived at through the subject's grasp of the "*esse* is *percipi*," not the other
way around.[16] Berkeley's mode of return to God, through the window
of subjective epistemological priority, contributes to the dependent and
shrinking status of a benevolent God who bestows a law-governed
world.

It is Hume who radicalizes the impulses in Berkeley's thought and
draws them toward the very skepticism the latter wished to avoid. For
if "nothing is ever present to the mind but perceptions," then "we can
never really advance a step beyond ourselves, nor can we perceive any
kind of existence but those perceptions."[17] Thus, for Hume, God's
power is as fallacious and inconceivable a notion as matter was for
Berkeley. At his most consistent, Hume argues that we must imagine
neither a world nor a constitutive God beneath our impressions: even a
substantial self becomes problematic.[18] The Humean mind experiences
an ongoing procession of atomistic impressions whose relations are always subject to doubt. In a sense, impressions are a kind of pure receptivity (though the "of what" is entirely inconceivable), but they are no
longer the reception of the meaningful world-gift of a generous God.
With Hume, we receive atomistic impressions whose relationships are
not given but are mere habits (e.g., causality) of a mind groping for a
sense that is always doubtful. We receive giftless contingency: Godless;
devoid of independent objects around which our impressions really cohere (again, where Hume is consistent); devoid of causality, power,
energy, and value. The pure receptive quality of impressions simultaneously short-circuits any substantial "that which" beyond the impressions received and cuts through any attempt to view impressions as
"given" by a willful subject. And so, this intransitive receiving leads to
an experience of the world as fraught with arbitrariness as was the
world of the pure subjective giving of "*esse* is *percipi*," which in the face
of a recalcitrant world was forced to deduce an (ultimately) no less
arbitrary God.

Kant as Epic Hero of the Secular

It is on this disintegrating terrain, where human selves are increasingly unable to be either the recipients of a being that is no longer the gift of a benevolent God or the self-originating givers of being itself, that Kant strides forth as the epic hero of the secular. He must come to terms with the crisis of receptivity and establish a self-given human sovereignty.

To understand the subtleties and problems of Kant's secular response to the crisis of receptivity so that we can move intelligently beyond it, we shall have to examine carefully the intricacies of his critical corpus. Yet to grasp initially some of the most essential aspects of Kant's project—what he thinks is at stake and the trajectory of the critical turn—there is perhaps no better place than the first and second prefaces of the *Critique of Pure Reason*,[19] where his project takes the form of an epic drama that powerfully exemplifies his secular aspirations. In the next section, I juxtapose with Kant's story a brief and idiosyncratic recounting of Nietzsche's *Thus Spoke Zarathustra* which introduces my notion of the "post-secular" in connection with the "gift-giving virtue." Nietzsche both illuminates the central ethical problems of the secular and (sometimes) gestures toward a post-secular *caritas*. My reading of *Zarathustra* helps articulate further the questions and concerns that infuse my engagements with Kant, Adorno, and Habermas in this book.

The tasks of the critical project are, Kant suggests, more epic and dramatic than all the epic dramas in history; and the accomplishments are more systematic, complete, and exhaustive. While his heroic efforts draw on and attempt to mirror the two most important revolutions of the past, the sublime magnitude of Kant's sense of his own work stems, on the one hand, from the extent to which he deems the critical project to be necessary to secure these past historical achievements and, on the other hand, from his sense that this project is the last truly grand philosophical-historical task. Kant's epic "Copernican revolution" consists in humankind's painstaking journey back to the secure homeland from its lost wanderings in a dangerous world of heteronomous objects to which humankind mistakenly believes its knowledge must conform. In contrast with the ancient epic, however, the homeland he seeks is none other than human beings themselves, as transcendental subjects, who give the conditions of knowledge and experience to which objects must conform. Safe within the subject of knowledge whose possessions have been submitted to a careful "inventory" and "systematically arranged," Kant enables humans to resist succumbing hopelessly to the seductive sirens who would mislead us into taking as our standard a "God's eye" view of things in themselves which we would somehow have to receive in order to have knowledge. Having made us safe within a subjectivity that gives a knowledge utterly free of wanderings with the ghosts of contingent alterity, Kant bestows on humanity a philosophical future unthreatened by "skeptical nomadism" and beyond all heroic deeds: a field of "amusement" in which "nothing can escape us" (Axx–xxi). The magnificence of Kant's epic effort grows in light of the "peculiar fate" with which he struggles, a fate that is unavoidable, "as prescribed by the very nature of reason itself," but also seemingly insurmountable, "calling humanity to tasks transcending all its powers" (Avii). Wrestling with humanity's intrinsic tendency to become lost outside itself, Kant faces self-destructive forces that can be defeated only through a steadfast gaze on the rational subject and its conditions of possibility, even as this very subject tends continually to move beyond itself in delusion. Kant's narrative of the transcendental subject's sovereignty is necessary as the form of therapy whereby humanity establishes that sovereignty which is its own self-forgotten essence.

Having never examined *itself*, reason takes the principles of experience to be a light shining from heteronomous things *themselves* and seeks to rise from this light toward the unconditioned light. In so doing, "human reason," as Kant graphically illustrates in his discussion of the antinomies, "precipitates itself into darkness and contradictions"; it becomes hopelessly lost in "the battlefield of the endless controversies . . . called metaphysics," the "scorned Queen of all the sciences" (Avii). In his (perhaps strange) analogical relationship with Copernicus, who wrote that "in the center of all rests the sun," Kant seeks to combine the idea of a light-giving center with that of an earth, from whose motion stems the apparent motion of all that surrounds it, into a narrative of subjective sovereignty that can cure humanity of the self-extinction of its own light.

In the absence of Kant's critical project the situation is grim: moving from "dogmatic despotism" to "intestine wars" to the "complete anarchy" of "the *skeptics*, [who were] a species of nomads, [who,] despising all settled modes of life, broke up from time to time all civil society." From this decay dogmatism reasserts itself, only to succumb again, repeatedly, until "the prevailing mood is that of weariness and complete *indifferentism*—the mother in all sciences of chaos and night" (Aix–x).

And this chaos and darkness does not remain confined to efforts to secure knowledge of that which lies beyond the limits of experience. Rather, its corruption is contagious, "showing itself in the midst of flourishing sciences, and affecting precisely those sciences, the knowledge of which, if attainable, we should least of all care to dispense with" (Ax). For Kant, Hume is the philosopher who most profoundly illustrates the pathways of contagion, by which nomadic skeptics and indifferentists creep beyond the body of the scorned queen, potentially to rot the entire terrain of human knowledge and experience. In attacking skeptically the rational a priori necessity of the relation between an effect and its cause, through which metaphysicians sought to proceed with regressive certainty to unconditioned knowledge, Hume confined human beings to a realm of experience (impressions) so narrow as to disintegrate everything: "David Hume came nearest to envisaging this problem, but was still very far from conceiving it with sufficient definiteness and universality. . . . He believed himself to have shown that . . . an *a priori* proposition [regarding the connection between cause and effect] is entirely impossible. If we accept his conclusions, then all we call metaphysics is a mere delusion" (B19). Moreover, such skepticism would threaten both the second epic intellectual revolution, that of natural science, and the first, that of mathematics, whose monumental significance Kant indicates when he writes that it was "far more important than the discovery of the passage round the celebrated Cape of Good Hope" (Bxi). Indeed, if Hume had realized the depth and generality of the problem in which he was entangled, "he would never have been guilty of this statement, so destructive of all pure philosophy. For he would then have recognized that, according to his own argument, pure mathematics, as certainly containing *a priori* synthetic propositions, would also not be possible; and from such an assertion his good sense would have saved him" (B20).

But this is not all that hinges on Kant's heroic deed, the critique of reason. "Above all . . . the most important task of philosophy" is to save morality and religion from the corrosive effects of dogmatism (which frequently espouses claims about freedom, God, and immortality that are incompatible with morality) and of a skepticism ultimately engendered by precritical metaphysics (Bxxxi). By rejecting the claims of metaphysics either to receive knowledge of the ultimate truths and conditions of things in themselves or to give itself theoretical knowledge beyond the limits of its proper extension, Kant purports to clear a space that the subject can later (in the Second Critique) "take occupation of" rightfully in the form of the postulates (freedom, God, immortality) that pure *practical* reason gives itself. While Kant is clear that "critique can never become popular," its effects are, he thinks, nevertheless historically far-ranging, making it possible "once [and] for all to prevent the scandal which, sooner or later, is sure to break out even among the masses, as a result of the disputes in which metaphysicians (and, as such, finally also the clergy) inevitably become involved to the consequent perversion of their teaching. Criticism alone can sever the root of *materialism, fatalism, atheism, free-thinking, fanaticism*, and *superstition*, which can be injurious universally; as well as *idealism* and *skepticism*, which are dangerous chiefly to the schools, and hardly allow of being handed on to the public" (Bxxxiv).

By describing the consequences that accompany the subject's self-forgetting reason, Kant (echoing his assessment of the securing of mathematics) shows his heroic deed to be "of such outstanding importance as to cause it to survive the tide of oblivion" (Bxi). In line with his understanding of the two previous intellectual revolutions, Kant seeks to discern not a world of heteronomous objects, but one in which "objects conform to our knowledge" (Bxvii). This effort to comprehend what the subject of knowledge (and, later, morality) has constituted according to a priori forms involves Kant in a painstaking analysis of the conditions of possibility of experience and, ultimately, an analysis of the unified sovereign subject itself. Central to this problematic is the question concerning how a priori synthetic judgments are possible; for it is through an exacting analysis of such generative judgments that the powers that produce the world of experience and the unified subject can be understood, along with the proper limits of those powers. By carefully recounting the extent and limits of our transcendental powers, Kant's narrative will enable us to utilize them to their fullest extent, without getting lost or overstepping their reach in self-destructive ways.

Kant's enactment and description of the heroic deed by which humanity might dispel its self-produced "chaos and darkness," so that light might break forth from a secure subjective sovereignty, is governed by two metaphors: the "trial" and the "inventory." Like natural scientists seeking that which they "have put into nature," by means of a trial in which "an appointed judge . . . compels the witnesses to answer questions he has himself formulated" (Bxiii), Kant seeks "to institute a tribunal which will assure to reason its lawful claims, and dismiss all groundless pretensions, not by despotic decrees, but in accordance with its own eternal and unalterable laws" (Axi). Kant does not believe that he can invent the law, but that he can secure it through a combination of exacting self-knowledge and a ruthless treatment as "contraband" of all which does not pass the test of self-given certainty: "it is not to be put up for sale even at the lowest price, but forthwith confiscated immediately upon detection" (Axv). Having established the character and limits of the subject's power to give the law, Kant then conducts an inventory to assess and further secure the subject's property. "In this field," he writes, "nothing can escape us" (Axx). On this terrain, carefully secured and assessed by means of Kant's epic narrative, the subject constitutes its sovereignty: the Sun King shines, is a sun, *precisely through* a critical turn in which, to paraphrase Marx, "man revolves about himself as his own sun."

We shall see that in the *Critique of Pure Reason* and the *Critique of Practical Reason* Kant sketches a series of strange and uncompelling efforts to postulate receptivity and otherness as mere functional correlates of sovereign subjectivity. This secular foundational narrative in which otherness (when functionalized) is extremely central, and at the same time (whenever its alterity might threaten sovereignty) persistently excluded, establishes a systematic violence and oblivion in the heart of the sovereign subject it is to secure. This tragic violence infiltrates and works at odds with the more admirable yearnings we find in the Kantian project. The analogy of the monological trial in which reason assures itself of its lawful claims is more than a little haunting in light of much that has happened over the past two hundred years.

Zarathustra's Awakening: Glimpses of a Post-Secular *Caritas*

Nietzsche takes Kant's solar sovereignty to extremes and conclusions that are terrifying. Yet he also illuminates the untenability and horror of this project, and points beyond it in promising directions with a power perhaps unrivaled in the nineteenth century. I proceed to a reading of Nietzsche's epic narrative—in order now to journey not toward a homeland of sovereignty but beyond this dangerous myth toward an agonistic giving and receiving, a post-secular *caritas* that moves beyond Kant's "I cannot and I must not receive" toward a tragic-but-not-hopeless "I must receive to give, but how can I?"

Thus Spoke Zarathustra[20] can be read as an exploration of the gift-giving virtue, the "highest virtue" (*TSZ*, 74).[21] What gift-giving is and how it can be is as much a question as an answer in Nietzsche's text. The reclusive saint who has retired from giving reveals early in the Prologue that people "are suspicious of hermits and do not believe that we come with gifts" (11). But the difficulty is not simply that the "others," the people, "the rabble," are not very receptive these days. More profoundly, the problem is enmeshed with something of which the old saint "has not yet heard . . . that *God is dead*" (12).

Yet in spite of and because of questions thus raised about radical contingency, power, heteronomy, and blindness, Nietzsche explores and seeks to affirm the gift-giving virtue. In part, Zarathustra journeys the harrowing paths of the gift-giving virtue because of his strong sense of the degrading and annihilating relations between selves that come to predominate where gift-giving is lacking. This negative motivation is cultivated through genealogies that aim to expose the illnesses that are intertwined with various modalities of the "sick-selfish" will to power: pity, selfish egoism, the state, equality mongering, neighborliness, last men, the marketplace, material acquisitiveness, ascetic selflessness, the jealous God of monotheism, the spirit of revenge and resentment. In each instance, Zarathustra perceives a weakening associated with the eclipse of generosity. He summarizes: "Tell me my brothers: what do we consider bad and worst of all? Is it not *degeneration*? And it is degeneration that we always infer where the gift-giving soul is lacking" (*TSZ*, 75).[22]

Yet what summons Zarathustra *toward* the gift-giving virtue as the condition of possibility of well-being is none other than the sun. The solar summons in the "Prologue" is borne on a powerful historical wave, spanning at least from Plato's solar analogy used to gesture toward *Agathon* (the Good, that which *gives* all beings being and perceptibility)[23] through Kant's Copernican revolution. Zarathustra experiences the sun as that which eternally overflows with a generous luminosity so graciously received by its earthy recipients.[24] This experience animates his often stumbling journey toward the gift-giving virtue:

"You great star, what would your happiness be had you not those for whom you shine?" (*TSZ*, 9–10). The sun that awakens Zarathustra, however, is not a *fundamentally* autonomous condition of possibility; rather, the sun is essentially entwined with those who receive its light. When Zarathustra exclaims "What would your happiness be had you not those for whom you shine?" we should recall that in *The Will to Power* Nietzsche defines happiness (pleasure) as the feeling of increasing strength and power.[25] In some sense the power and being of the sun is connected with others, as Zarathustra explicitly notes when he says, "You would have tired of your light and of the journey had it not been for me and my eagle and my serpent" (*TSZ*, 9).

Insofar as the *condition* of the power and being of giving involves receivers, the giving-ground is pierced with contingencies of possibility and danger that erode pretensions to independent sovereignty. In search of receivers, Zarathustra is drawn down from his cave toward tremendously difficult questions. For despite his telling inability to receive the reclusive saint's warning about the extreme difficulties of being received, Zarathustra soon discovers the recalcitrance that meets his giving. And at the deepest and highest levels Zarathustra is challenged to question his understanding not only of the recipients but of gift-giving itself. This in turn forces Zarathustra to radicalize the entwinement of giving and receiving, ultimately pushing him beyond his opening formulations of solarity. A theory of receptive generosity as the wellspring of intelligence, power, and well-being gradually emerges through the relatively small fissure of receptivity in the "Prologue."[26]

Zarathustra's first encounters with people in the marketplace go exactly as the old saint said they would: Zarathustra's efforts to give are smashed on the shores of those unwilling to receive him. In turn, Zarathustra receives not receivers but a corpse. But is it the tenacious stupidity of the herd which alone bears responsibility for these disastrous encounters? Or should one also include the blindness of the solarity that governs Zarathustra's giving? If the latter, Zarathustra seems to have little clue. In his speech on the "gift-giving virtue" at the end of part one, he still resists advancing to the question of the other *as other* and locates the origin of this virtue in being "above praise and blame," where "your will wants to command all things" (*TSZ*, 76), perhaps not unlike Kant's monological tribunal.

Yet Nietzsche traces Zarathustra's solar wanderings throughout part two in parables that "do not define, they merely hint" (*TSZ*, 75), and they do so in ways which increasingly bring to the fore the mounting tragedies and weaknesses accompanying this position. Significantly, Zarathustra, blinded by the sun he seeks to emulate, is incapable of such self-reflection until near the end. Part two opens with him startled awake by a dream: a child holds a mirror before him in which "it was not myself I saw, but a devil's grimace and scornful laughter" (83). He takes this as a sign that his teaching is endangered, that his gifts are giftless, failing. They are. Before such an image one might expect some reflection on possible ways in which Zarathustra might be implicated in these dangers and failures. Yet the solar blindness, which ruins the very giving it guides, simultaneously blinds him to such reflection and instantly he externalizes the problem: "My enemies have grown powerful and distorted my teaching till those dearest to me must be ashamed of the gifts I gave them" (83). Secure in this account, he leaps up and "like dawn" proclaims that he will once again go down to his friends and enemies, giving like a plunging river of love: "Mouth have I become through and through" (84). But is this not precisely an "inverse cripple" (138), having developed one organ to the detriment of all else? Mouth he is! But can he see, hear, touch? Can a mouth alone be radiant? Giving?

It is not long before Zarathustra turns his mouth on "the rabble," uttering terms remarkably similar to those he used to describe the image of *himself* in the mirror, and once again he so proclaims without a moment's reflection on the semblance. Echoing the "devil's grimace and scornful laughter" of his own image, he speaks of the image of the rabble mirrored in the well they poison: "grinning snouts" and "revolting smiles" (*TSZ*, 96). Are "the rabble" closer than he thinks, peering from out of his own sun? Zarathustra rages and fumes against the rabble, and he closes his speech saying, "Like a wind I yet want to blow among them one day, and with my spirit take the breath of their spirit" (99). Is it radiance, power, and giving we hear here? Or something else?

Again with pointed irony, Nietzsche opens the section that follows Zarathustra's wind fantasy with a parable of storm-provoking tarantulas whose "poison makes the soul *whirl* with revenge" (*TSZ*, 99). Of course, Zarathustra immediately construes these spiders in a wholly external way: they are the type exemplified by the punishing equality police and courts. He lets them close enough to admit that he has been bitten; but he leaves us with a strong sense that he has risen above the poison, for "Zarathustra is no cyclone or whirlwind" (102).

Once again, though, the question concerning whether he has been bitten serves to conceal. For lurking in the nagging ironic background that Nietzsche provides for Zarathustra are deeper questions the latter avoids: Is there a tarantula hiding in his (not quite) sovereign sun? Has he bitten himself? Is the rabble poisoning the well partly a manifestation of his own "sun-poisoning"—in addition to all that he does identify? And what might it be about solarity that could make it so? The "Night Song" parable which soon follows revolves around these questions.[27]

From the depths of darkness, Zarathustra exclaims: "Light am I; ah, that I were night!" (*TSZ*, 105). As a ceaseless and self-originating giver of light, he is unable to receive anything. "Many suns revolve in the void: to all that is dark they speak with their light—to me they are silent. Oh, the enmity of light against what shines: merciless it moves in its orbit" (106–7). Significantly, he says: "I do not know the happiness of those who receive" (106). Recalling here Nietzsche's understanding of happiness as a feeling of increasing power, Zarathustra's exclamation "Oh, darkening of my sun!" gestures toward the self-defeating character of solarity. It would appear that the very power necessary for one to be as an overflowing of giving would require one to receive more than *others-as-receivers*. What is yearned for here—what seems necessary for radiant generosity and for power itself—seems to be the capacity to receive partially the other *as other*, as another light, another voice. In the absence of this: "My happiness in giving died in giving; my virtue tired of itself" (106).

Why this weakening, tiring, and darkening? Could it be that the unreceptive giver—no longer either creating or receiving the stable ontological ground that sufficiently guides one's relations with others but, rather, being drawn together and pulled apart in the context of agonistic and incomplete contingent identities—becomes incapable of cultivating a gift, incapable of experiencing those wild yet more receptive and *discerning* dialogical encounters with the often chaotic otherness of the world which are necessary for the emergence of intelligence, for the birth of a "dancing star"?[28] Could it be that the height of the highest virtue is attainable only through more powerfully receptive relations with others *as others*? How could Zarathustra hope to give to those of whom he knows nothing, from whom he has received so little? "They receive from me, but do I touch their souls? There is a cleft between giving and receiving; and the narrowest cleft is the last to be bridged." For "the heart and hand of those who always mete out become callous from always meting out" (*TSZ*, 106). They lose all sense of the other, all orientation concerning what might be empowering, what might shame the other.

Yet if Zarathustra's agony in "Night Song" brilliantly opens onto this wisdom, it soon disappears again in blinding flashes of the solarity by which he is seized. Unable to receive, unable to give, Zarathustra's giving turns unpalatably sour. And a giving whose fundamental structure dooms it to failure leads to resentment. "I should like to hurt those for whom I shine . . . rob those to whom I give. . . . Such revenge my fullness plots: such spite wells out of my loneliness" (*TSZ*, 106). But has he not, then, become the revengeful tarantula he so despises—this spider who dwells where the sun shines brightest and hottest?

The wisdom of "Night Song" has an intermittent presence in the story of Zarathustra, and sometimes it is entirely eclipsed. Yet Zarathustra returns to it at numerous key points in his journey, reaching higher and deeper in time. In the section "On Redemption" Zarathustra seeks to embrace *difficult* reception as vital and life-giving, like the kind that he suggests the hunchback who would rather be "cured" ought receptively to embrace: "When one takes away the hump from the hunchback one takes away his spirit" (*TSZ*, 137). Still, Zarathustra recognizes a problem within the solar will that makes reception itself difficult, nay impossible, to embrace. For the passive aspect of our relation to time is ineliminable and gives the lie to the will's claims to be self-originary giving. The present moment of the will is carried along by an intractable past which is more than its will and which cannot be changed "at will." "The now and the past on earth—alas, my friends, that is what *I* find most unendurable" (138). Facing the past, the self-proclaimed unreceptive will seems impotent and mythical. "The will is still a prisoner. . . . 'It was'—that is the name of the will's gnashing of teeth and most secret melancholy. Powerless against all that has been done, he is an angry spectator of all that is past." Angry at that to which the will must unwillingly receive, the solar will becomes a destructive force and "wreaks revenge," which Zarathustra defines precisely as "the will's ill-will against time and its 'it was.'" Unless the will can receive otherness in the fundamental form of temporality, "cloud upon cloud rolls over the spirit": the sun extinguishes itself (140).

But how to receive time, through which the other and otherness have come and are always already coming? Somehow the will must receive time, gather together the "fragment, riddle, dreadful accident" which temporality appears to be, and say "Thus I will it." But how, when our history is permeated by so much rabblishness (and also the brilliance of others that one "girt with light" finds difficult to perceive/ receive)? Nietzsche seeks a sort of redemption in the face of recalcitrant time and rabblishness through his idea of the "eternal return." Whether or not Nietzsche really thought this doctrine had literal ontological merit, clearly he pondered it as a practical regulative idea (an idea he called "the greatest weight");[29] and it is this that I wish briefly to sketch in the context of the question of receptivity.

Zarathustra's animals articulate the core of the idea: "All things recur eternally, and we ourselves too; . . . You teach that there is a great year of becoming, a monster of a year, which must like an hour glass turn over again and again. . . . [A]ll these years are alike in what is greatest and what is smallest; and we ourselves are alike in every great year, in what is greatest as in what is smallest. . . . [T]he knot of causes in which I am entangled recurs and will create me again" (*TSZ*, 228). Much of Zarathustra's effort focuses on coming to terms with the implications of this thought: "The eternal recurrence even of the smallest . . . that was my disgust with all existence" (219).

One cannot escape nor will away the smallest in others and in oneself any more than one can the past. The question, then, which is pressed into being and opened under the weight of this highly pressurized thought of passive receptivity, is how to receive this smallness (and grandness) in such a way that radiance and giving do not darken but are instead made possible in part precisely through this reception. There is no singularly triumphant answer to this question, despite moments in the text where joyful triumph seems absolute. For distance, opacity, difference, and rabblishness, which are in part the space of giving's possibility, simultaneously permeate giving with tragic dimensions of error. Instead, eternal recurrence, the thought of unending *closed* time, hangs over Zarathustra in an *essentially* interrogative aspect—as a question through which the *opening* of time as a site of possibility for the creation/coming of the higher emerges. The interrogative overture (a word capturing the essential connection between opening and height) is endlessly renewed in the question of how one might receive the rabblishness (and grandness) within and without in order that it might be gathered together into a giving and a gift high enough to redeem it, high enough to say yes to the eternal return of this moment. The question involves a partly agonistic, partly cooperative—always transfiguring—dialogical effort with others to discern what is lower and what is higher; to discern how these differences and distances might be brought together and held apart such that we might become more receptive of their gifts, more capable of giving, less resentful and revenge-seeking, more radiant. This entwinement of giving and receiving is the precarious elaborating foundation of well-being and sense.

The gift-giving virtue is what is highest. Yet its greatest possibility for emergence "arises," paradoxically, underneath "the greatest weight"— the thought of eternal return (which is hence the greatest gift?). The greatest weight presses us generously into the *depths* of our surroundings as the oblique path of ascension. Hence Zarathustra says to himself that, in contrast to those who are "obtrusive with [their] eyes" and who are stuck to the "foreground" surface of things, "you, O Zarathustra, wanted to see the ground and background of all things," wanted to plunge into the depths of beings, into those aspects and possibilities which are concealed beneath immediate appearances (*TSZ*, 153). "It is with man as it is with the tree. The more he aspires to the height and light, the more strongly do his roots strive earthward, downward, into the dark, the deep—into evil" (42). Into evil because the background depth of beings is barred from generous approach by the taboos of evil (races, sexualities, classes, practices, desires, thoughts, bodily expressions, and so forth). Yet, in the stream of Nietzsche's thinking that I'm following here, the striving into evil is to be animated and circumscribed by that generous respect for otherness which is solicited by the highest virtue. It is through this agonistic giving and receiving in *depth* that one can best affirm life and possibly *rise* toward a joy capable of dancing in the face of the eternal question of the eternal return. A suppleness and interrogative comportment is vital to the virtues Nietzsche would have us pursue. Hence it is fitting that Zarathustra appears to reject the idea of eternal return articulated by the animals; for if it were to become a doctrine and not a question-engendering question, it would cease to animate the very receptivity strived for. The ideal always partly suffers in its articulation and incarnation, even as it must be incarnated. Its *logos* is always "wounded." Finally, therefore, receptive generosity remains a soliciting ideal whose realization is "not yet." It is a direction toward which we bring forth children, a direction from which they are coming. Like Zarathustra's children, who shall be "taciturn even when [they] speak, and yielding so that in giving [they] receive," these children shall be capable of friendship (161). These children are still coming to be on the final page: near, but not yet here.

In what follows I explore possibilities for articulating, soliciting, and (to a much lesser extent) practicing an ethic of receptive generosity. It is my sense that the death of at least a *sovereign* God and subjectivity constitutes a desirable space for such activity, rather than a necessary site of ethical dissipation. At a time when generosity often appears to be poorly construed, or a cover for greed, or an outdated dream in this new sleep called "the end of history"—at such a moment, these reflections are at once timely and untimely. Be that as it may: this bottle to the sea.

CHAPTER ONE

Kant: Secular Sovereignty and Beyond

In this chapter I elaborate my understanding of Kant's theory of secular subjectivity. The first two sections trace key maneuvers in Kant's sovereignty narrative in the epistemology of the *Critique of Pure Reason* and the morality of the *Critique of Practical Reason*. In the third section, I argue that this narrative orients Kant's thinking in ways that involve a radical eclipsing of the other. Then, in a lengthy concluding section, I develop provocative chapters in the *Critique of Judgment* that point beyond secular sovereignty toward a post-secular terrain of agonistic giving and receiving.

Lawgivers and Receptivity in the First Critique

Within the context sketched out in the Introduction—i.e., the untenability of the early-modern self, as either a recipient of being or a self-originating giver of being itself—Kant awakens from his troubled slumber and formulates a philosophy in which humans secure their sovereignty as "law-givers of nature" (*die Gesetzgebung für die Natur*)[1] and lawgiving moral wills. Yet the distinct power and revealing equivocations of this position stem from Kant's insight that the "transcendental story" of our lawgiving sovereignty must still involve central dimensions of receptivity.[2] Pure receptionless giving can lead only to arbitrary *sui generis* subjectivity or to an equally arbitrary and opaque subjective necessity "implanted in us" by an omnipotent God (B168). In short, "were I to think an understanding which is itself intuitive (as, for example, a divine understanding which should not represent to itself given objects, but through whose representation the objects should themselves be given or produced), the categories would have no meaning whatsoever in respect of such a mode of knowledge" (B145). They would not exist as universal and necessary conditions of possibility for experience, but merely as the whimsical forms of experience given by a subjectivity capable of limitless change. The objective universality and necessity that Kant thinks we must give to the world and the moral realm, if we are to be sovereign, is conceivable only in a context that has receptivity at its core: Kant must reimagine, in the absence of a foundational generous God, an *intertwining* of giving and receiving between self and otherness that can secure the subject's stable, lawgiving sovereignty. However, this otherness (and receptivity itself) must be purged of the disruptive contingency brought into sharp focus by Hume. It must be thinned out to the point of absence and yet present at the same time: a functional presence-absence for the sovereign subject and its transcendental story. Like Hegel's master, Kant's sovereign subject requires an otherness which is not really otherness, and thus it must be narrated through equivocations in which it is slipped in and out of being in order to fulfill the functional requirements of sovereignty.

Central to this project is the move away from transcendental realism, which "treats *mere representations* as things in themselves" (A490–91, B518–19), to transcendental idealism, "the doctrine that appearances are to be regarded as being, one and all, representations only, not things in themselves, and that time and space are therefore only sensible forms of our intuition" (A369). Since Allison has very lucidly developed the centrality of this position for understanding the First Critique, I shall draw substantially on his discussion.[3]

Though it might appear that the above formulation sounds a lot like Berkeley, as Allison notes, Kant finds Berkeley and Hume to be as firmly planted in transcendental realism as are Descartes or Newton. Both of the former "treat . . . impressions as if they were given to the mind as they are in themselves. . . . [T]hey regard appearances (in the empirical sense) as if they were things in themselves (in the transcendental sense)."[4] In other words, though objects depend on the mind for the two empiricists, they do not "conform to our knowledge," as Kant stated in his formulation of the Copernican revolution, but appear as a kind of brute given data. Insofar as knowledge mediates and relates diverse impressions, Berkeley's and Hume's perceptions stand outside knowledge in their utter particularity and simplicity. Questions of law, order, and continuity are external to brute perception. (Hence Berkeley introduces God, while Hume becomes skeptical.) On Kant's reading, Humean perceptions are ultimately taken as the "real" which can be only guessed at probabilistically. In contrast, for Kant, the categories of our knowledge and even space and time become conditions of possibility for any experience, without which we can have no cognitive apprehension of objects. Thus, the fundamental heterogeneity of knowledge and experience disappears for Kant.

At the same time, though, a new heterogeneity emerges. For if all experiences must conform to what Allison aptly calls "epistemic conditions," then "things in themselves" lose the presence they had as brute data for the transcendental realists; things are stripped of all qualities of human knowledge and experience and become the idea of shear cognitive inaccessibility. Yet this understanding of epistemically conditioned appearances and their radically inaccessible correlates, far from simply banishing the idea of things considered in themselves to the realm of the absolutely unspeakable and irrelevant, in fact bestows on objects considered as things in themselves a special importance which incites speech and is central to Kant's "transcendental story."

This importance is manifold. Kant's distinction between what we can "know" (as the ability to prove a thing's real experiential or practical possibility) and what we can "think" (as that which is merely not logically self-contradictory) *allows* Kant to think and thus speak of things considered in themselves.[5] And the logical-semantic connection between appearances and things in themselves *demands* the thought of the latter, lest "we should be landed in the absurd conclusion that there can be an appearance without anything that appears"(Bxxvi–xxvii). Moreover, Kant emphasizes the negative transcendental imperative to "mark the limits of our sensible knowledge" with the thought of things in themselves so that we do not overextend it in ways that lead to the devastating antinomies (A288–89, B344–45). Yet while these considerations are very significant, even more important is that this absolute otherness must become a central figure that both makes conceivable and helps establish a decontaminated and pacified receptivity in relation to which the autonomous subject can become a secure lawgiving sovereign. It is not just the *limits* but, moreover, the *interior* of this sovereignty and the nature to which it legislates that is defined by this consideration of objects as things in themselves.[6]

Kant needs otherness in order to avoid the arbitrariness of a pure self-giving subject, but receptivity appears to confront him with a heteronomy that is equally disruptive of a sovereignty of objective necessity. Kant, then, must construe receptivity and otherness so that they are drained of their disruptive contingency. The successful negotiation of this problem hinges on a critical examination of "receptivity."

For Kant, the essence of receptivity is passivity. This passivity constitutes the manner in which we must consider the relation between sensibility and things in themselves, on the one hand, and the character of what it is that sensibility gives the understanding for the understanding's peculiar synthetic activity, on the other. If the transcendental narrative did not begin with the thought of a passive-receptive relation to otherness, then the very passivity of sensibility which gives the sensible manifold that character so seemingly suited to the active subject's legislation of objective necessity would be unimaginable.

Throughout the First Critique, the "two fundamental sources" of knowledge, sensibility and understanding, are contrasted respectively in terms of receptivity and spontaneity. Spontaneity is the understanding's activity of combining, determining, and uniting the manifold of spatial and temporal intuitions given in sensibility, the "passive" faculty of the subject (B153),[7] in order to bring it under the unity of apperception (i.e., a single self-consciousness). The passivity of the receptive faculty is definitive of the relation between sensibility and things in themselves discussed in the Transcendental Aesthetic. This passivity is hinted at in the opening words of the narrative, where the repetitive employment of the indefinite and perhaps even perplexed adjective *welche* ("In whatever manner and by whatever means a mode of knowledge may relate to objects" [A19]) suggests a relation that simply "happens" to sensibility, a relation in which it is too passive to be able to grasp what happens.[8] We learn immediately that this relation is not merely something that *contingently happens to* sensibility but, rather, *that which defines its being*: "The capacity (receptivity) for receiving representations through the mode in which we are affected by objects, is entitled *sensibility*" (A19). The idea of unknowable things in themselves and that of passive receptivity engender each other. Sensibility's passivity guarantees the unknowability both of things in themselves and of our relation to them, while unknowable things in themselves enable Kant to imagine a sensibility that is truly passive and receptive in the midst of all that happens *to* and *through* it: given sensations, it seems, are "the effect of an object" (A19–20), not the object's own activity.

This first critical insight is connected to a second one which is equally central to the transcendental story. The passive receptivity of sensibility profoundly colors the character of the manifold of sensations and pure intuitions that are "given" through it. The effect of things in themselves on our *receptive* sensibility "*gives*" objects of intuition, but these "objects" are "*undetermined (unbestimmte) objects*" (B34, A20) be- cause they are given through a passive receptivity which, as passive, cannot make or hold together anything determinate. (Kant repeatedly refers to the manifold as entirely uncombined [B129ff.].) Hence, Kant writes, that the form of sensibility (or appearance) does not itself create determinate form, but "so determines [*macht dass*] the manifold of ap- pearance that it *allows of* being ordered [*geordnet werden kann*] in certain relations" (A20, B34, my emphasis). The theme of a "determining" which "allows," but does not itself accomplish, a determinate order of determinate things is reemphasized when Kant writes of sensibility as a form in which "sensations *can be* posited and ordered in a certain form" (A20, B34, my emphasis): the positing must be considered an activity of the object on our receptive faculty, the ordering an activity of the un- derstanding.[9]

Hence we see that for Kant the rigorous notion of passive receptivity requires us (a) to think of things in themselves which effect sensibility and (b) to understand that which sensibility receives from things in themselves as indeterminate. The transcendental narrative, then, must consider things as absolutely other—apart from space, time, and the categories of our understanding—but this otherness is by definition that which is permanently closed to all other interrogative considera- tions. Radical otherness plays a vital role as a functional correlate of our sovereignty. For radical otherness makes possible the conception of sensibility-as-passive-receptivity, which gives a manifold of indetermi- nate representations that, as we shall see, are required by and allow the rule-governed synthesis of understanding and imagination. Thus Kant believes he has surmounted the problem Hume faced in the idea of a determinate heteronomous receptivity that disrupts understanding. The concept of reception is transfigured and secured by considerations of things in themselves and by the indeterminacy of receptivity. For Kant, then, there is both a more radical thought of otherness than there was for Hume (things in themselves) and simultaneously a purging and dissolving of otherness (in the indeterminacy of reception). Both the absolute unknowability of things considered in themselves and the indeterminacy of their effects on sensibility ensure an account of the other that is devoid of anything disruptive of our lawgiving sovereignty.

Yet my discussion of receptive sensibility and indeterminacy makes use of a series of active (and even pejorative) verbs seemingly ill-suited to describe the absolute and ostensibly innocent passivity (through which all combination and determinacy are impossible) that I first attributed to Kant's theory of sensibility. This tension calls for further exploration. That the usual notion of passivity (as inactivity) does not exhaust Kant's theory of sensibility or intuition is indicated first when he refers to sensibility as a capacity (*Fähigheit*) or power (*Vermögen*). This seems paradoxical: "It seems to be stretching the senses of *Vermögen* and *Fähigheiten* quite thin to say that receiving impressions is something we have the ability to *do*."[10] But it is precisely by clarifying this confusing terminology that we can more fully grasp the notion of sensibility and illuminate the activity of receptivity as *pacifying*.

"The capacity [*Fähigheit*] (receptivity) for receiving representations through the mode [*Art*] in which we are affected by objects, is entitled *sensibility*. Objects are given [*gegeben*] by means of sensibility, and it alone yields [*liefert*: provides, supplies, delivers, produces] intuitions" (A19, B34). In this portrayal of sensibility, the passive connotations of receptivity are juxtaposed with the more active language of "capacity"—that "by *means* of" which objects are "given"; that which "alone yields us intuitions"—all of which evokes a way of *doing* something.

Yet clearly this means of giving and supplying is not giving in the same sense that understanding is giving when, in giving the law to nature, it unites, combines, and determines a manifold of representations. Rather, as we have seen, the activity of pure receptivity must be grasped as a giving which yields indeterminate representations and intuitions. It is in this sense a "determining which allows" of being ordered by a synthetic faculty. While my initial gloss emphasized receptivity's double-faced passivity—with respect to the affection of things considered in themselves and with respect to the activity of the understanding—it is equally important to grasp that the translation of receptivity vis-à-vis things in themselves into a receptive allowing vis-à-vis the synthetic faculty requires a "making indeterminate" which involves more than receptivity understood as inactive passivity.[11] Rather, the hinge on which the face of passive receptivity turns from things themselves to the understanding, requires the notion of sensibility as a *pacifying* of the effects of objects, such that given intuitions are purged of all recalcitrant heteronomy vis-à-vis the understanding. For Kant, the extreme of the passive is that which *maintains* its passivity through the activity of pacifying: *an active passivity that pacifies.*

This claim is made plausible initially by considering the possibility of inactive passivity. It is not clear what *inactive* passivity could give us. A receptive passivity that is somehow analogous to a material which is impacted on and set in motion (as in Hobbes's account), or "stamped" (a trans*forming* record of unknown affects), will not do, for this would imply that sensibility gives a determination of some sort that would be prior and recalcitrant to understanding. Inactive passivity *might* be conceived as analogous to a form totally devoid of energy. But in this case it would seem to allow the dissolution of all into an abyss and give nothing. In short, as inactive passivity, sensibility would be too determinate or too indeterminate to allow the narrative of transcendental sovereignty to get off the ground. However, in Kant's story, the *active* passivity of sensibility yields not indeterminate indeterminacy but the specific (though, as we shall see, specifically *equivocal*) indeterminacy of a *manifold* that allows itself to be ordered by the understanding. I think this pacifying is what Kant has in mind when he ascribes a "synopsis" to the manifold of intuition.

Kant describes the manifold of intuitions given by a pacifying sensibility as a "homogeneous manifold" of "points" (A162–63, B203) of sensation (i.e., pure intuition, lacking sensation), or "representations" [*Vorstellungen*] (B129), that occupy only an "instant" or "eyeblink" (*Augenblick*) of "no extensive magnitude" (A167, B209). Each of these representations or points must be thought of as utterly uncombined and thus as having no temporal or spatial extension, since all extension has parts that have been combined. They must be utterly simple, indivisible, lacking internal relation. Kant claims that without a synthesis that recognizes, holds together, and reproduces for a unitary consciousness the manifold of parts, "not even the purest most elementary representations of space and time could arise" (A102). Hence the manifold must contain representations that are really only space and time *in potentia*.

This brief description begins to illuminate my claim that receptivity is a "pacifying that allows." It appears that receptivity, as a pacifying activity, determines the effects of things in themselves so as at once to dissolve relations and offer homogeneous, indivisible, eyeblink-like points. In this way, Kant seeks to avoid the abyssal and arbitrary possibilities that appear to accompany the indeterminate indeterminacy of an inactive reception of the effects of otherness. By dissolving all relations, Kant hopes to avoid the specter of a heteronomy before which the understanding would be arbitrary. Simultaneously he avoids the abyss, which would supply nothing, by offering the idea of a homogeneous manifold of pacified, extensionless *Augenblicken* whose essence is to "allow of being ordered" by the understanding. Sensibility receives the effects of the object in itself in a manner that gives the understanding (as the lawgiver of nature) what it needs: (a) a manifold form and content which is at once given independently and thus demands a rule-governed synthesis in order to be brought under the unity of apperception; (b) a manifold which is simultaneously drained of any determinacy and prepared for understanding's diverse synthetic activities. Indeed, the given manifold both allows and demands the sovereign rule-governing activity of the imagination and the understanding without which experience and the self would dissolve.[12]

Yet the idea of an otherness that is *nothing but* a functional condition for the sovereignty of the subject is, as Hegel so sharply perceived, not the idea of an other at all.[13] And this absurdity of positing an other-not-other undermines the tenability of Kant's narrative of sovereignty just as surely as it does the master's search for recognition through a slave in Hegel's dialectic.

To elaborate this claim, we must examine Kant's description of the relation between sensibility (or intuition) and understanding.[14] As shown, pure sensibility presents a formally homogeneous manifold of spatially and temporally unrelated representations. Understanding combines this manifold in a spontaneous act of synthesis, the latter term "indicating that we cannot represent to ourselves anything as combined in the object [or intuition] which we have not ourselves previously combined" (B130). Kant depicts synthesis as an activity that can be discussed in three dimensions: apprehension in intuition, reproduction in imagination, and recognition in a concept. On the synthesis of apprehension, Kant writes: "In order that unity of intuition may arise out of this manifold (as is required in the representation of space) it must first be run through and held together" (A99). Here is the original German and my own, perhaps more helpful translation: *So ist erstlich das Durchlaufen der Mannigfaltigkeit und dann die Zusammen-nehmung desselben notwendig* (It is first the running through of the great diversity and then the taking or holding together of the same that is necessary). With synthetic apprehension, Kant begins to formulate the conditions of that absolute spatial unity which belongs to a "single moment" (*Augenblick*). This synthesis is "inseparably bound up with the synthesis of reproduction," which reproduces each part ("part" here means an *Augenblick* of potential time or potential space) as the coming parts are run through and which holds them together. "Obviously, the various manifold representations that are involved must be apprehended by me in thought one after another. But if I were always to drop out of thought the preceding representations (the first parts of the line, the antecedent parts of the time period . . .) and did not reproduce them while advancing to those that follow . . . not even the purest most elementary representations of space and time, could arise" (A102). This synthesis of reproduction, a "transcendental faculty of imagination" (A102), is entwined with a "synthesis of recognition in a concept" that generates a unity "in accordance with a rule" such that the unity of apperception, the singular "I think," recognizes that which is being synthesized as necessarily belonging together (A104–5) or belonging together according to necessary relations. Such a categorical synthesis generates the concept of an "object $= x$" in which all experience is united. Hence Kant presents us with an idea of extensionless parts being "run through and held together" by an activity of reproduction and recognition which ultimately unites the manifold in necessary categorical relations, quantities, qualities, and modalities through which the world of experience fundamentally coheres. This rule-governed synthetic activity "gives the law" to nature by combining the manifold into experienceable objects: it gives, makes, produces the categorical form of nature.

Yet this narrative of subject-generated coherence is itself undermined by the incoherence of functional equivocations. For what are we to make of a statement like "The manifold of representations can be given in an intuition which is purely sensible, that is, nothing but receptivity" (B129), which opens the second-edition Transcendental Deduction? How are we to think of these "representations" or "parts"?

They must have some substantiality, because they are the "that which" synthesis acts on. No matter how Kant understands synthesis, we must imagine something "given" which is to be synthesized. Otherwise, combination becomes as meaningless as, for example, receptivity of nothing. It would yield nothing. Furthermore, we must imagine something given because it is precisely the manifold uncombined character of this given, these "representations" (*Vorstellungen* literally, "placed before" or "before-placed"), that both *necessitates* and *allows* the synthetic constitution of experience. Because there can be no coherent self or world without the law-governed connection of these representations, we can understand Kant's claim about the lawful necessity of nature and its formal origination in the lawgiving subject. The necessity of law requires the idea of something given which our rule-governed synthesis must combine. These representations must be uncombined in relation to one another and must lack all internal combination, for any combination would confront the synthesizing subject as an *other* recalcitrant order, before which synthesis would appear arbitrary.

Yet, from an epistemic point of view, this required idea of a given without any extension or combination is precisely the idea of *nothing*. Of course, Kant acknowledges that without transcendental synthesis, "it would be possible for appearances to crowd around the soul, and yet be such as would never allow of experience . . . and consequently would be for us as good as nothing" (A111). They would be "merely a blind play of representations, less even than a dream" (A112). My claim, however, is that the manifold cannot even *be* an unexperienced crowd or a *blind* play of representations. Each "representation" in the manifold must, as utterly uncombined, have *no* spatial or temporal extension; for all extension is divisible, and divisibility implies a prior synthesis. At most, these representations can be sensations or *potentia* for sensations of zero-extensive magnitude. Such a notion is utterly unintelligible; and it is equally unintelligible how extension could emerge through the synthesis of such "parts." And even if extension *could* somehow emerge, it would require an entirely different type of synthesis than we find in the sovereign lawgiver narrative concerning "parts" which are run through and held together categorically. Kant's synthesis requires a *given* manifold to act on. Yet because all synthesis occurs in the imagination and understanding, he must simultaneously conceive of this manifold as nothing: extensionless and nonrecalcitrant. Kant's exclusive understanding of synthesis requires us to imagine a manifold that is less than that which is required by the idea of synthesis as *acting on something*. In the face of this problematic equivocation, the rational subject's narrative of lawgiving sovereignty trembles at the edge of meaninglessness.

Shifting the lens of our analysis, let's imagine that somehow something *is* given: "representations" which, though "less than a dream" and prior to being determinately "placed" through synthesis, are nevertheless sufficiently "given" to "crowd around" in a "blind play." Numerous interpretations lean toward this more substantial view of the constituents of the manifold, a view that indeed seems necessary if "run through and held together," "reproduced," "combined," etc. are to make any sense. With Prichard, then, Kant might mean that sensibility gives "isolated data of sense," "units";[15] or, with Ewing, "individual existents," "presumably small durations and extensions."[16] We might contend, with Paton, that Kant didn't mean we are given absolute unities but was just expressing "a limit reached by analysis."[17] Or, following Wolff, that the manifold consists of diverse simultaneities we cannot *know*.[18] Or is Allison right when he asserts that the question whether we are provided particulars or organized givens is not as important as the idea that, with the manifold, we are given a "proleptical" awareness of representations?[19] Although many interpretations appear to fly in the face of Kant's claims about givens that have "no extensive magnitude," are "absolute unities," "uncombined," and so forth, they merely make explicit what we must imagine if we are to avoid absurdity when we use terms like "given," "part," "data," and so forth. Insofar as there are "small durations and extensions," or for that matter *anything* given to the understanding by our receptive faculty, we must presuppose some synthesis or at least "synoptic" organization prior to the threefold one discussed above. And these orders—which even if utterly lacking in relation to one another nevertheless bear relations within—must present the imagination and understanding with an impenetrable heterogeneity that they did not themselves produce.[20] Consequently, these parts must be black-hole-like presences for the unity of apperception, or at best indeterminate stuff we merely guess at, and it is difficult to imagine a Kantian nature synthesized out of such things (or a self whose sovereignty would not dissolve in their midst).

In short, if the thought of the received manifold as extensionless always offers the synthetic power *too little* for the transcendental story, then the thought of "whatever something" always gives *too much*. We shall see the idea that the nonidentical always says too little and too much with respect to our concepts (and vice versa) at the heart of Adorno's negative dialectic. But that dialectic was animated by a notion of receptive generosity at once far grander and less hubristic than that inscribed in the idea of a sovereign lawgiver of nature. For Kant, nonidentity is the death, not the life, of the self.

What we witness here, I think, are the contradictions that emerge as Kant attempts to narrate the subject's sovereignty in relation to an otherness-not-otherness. First of all, we must think of an otherness "in itself" in a manner wholly subordinated to the requirements of a tale of the subject's objective sovereignty. This "thing considered as in itself," however, is never so considered. Rather, it is an "in itself for us" insofar as we speak of its existence only in a manner that establishes a pacifying receptivity which gives the understanding a manifold that demands lawgiving and is devoid of recalcitrance. Next, Kant requires an otherness given by and through our receptivity (if lawgiving synthesis is to have any matter and meaning). Yet he requires just as surely that the otherness be nothing—i.e., utterly devoid of extension or order—lest it undermine the active synthesizing subject with impenetrable heterogeneity.

The imagination of an otherness which is an absolute extensionless unity is not functional for Kant because of some specific coherence as an idea. For not only can't we *know* what this otherness would mean, it is moreover *unthinkable*, since "given," "data," "part," "representation," "perception," etc. are in *contradiction* with the predicates "absolute unity" and "lacking extension and duration."[21] Rather, this extensionless indivisible otherness is functional because it evokes a very specific incoherence, an antinomial oscillation in the mind,[22] between being and nothingness, that can be deployed now this way and now that, depending on whether understanding needs the presence of an otherness, on the one hand, or nothing recalcitrant and heterogeneous to itself, on the other. Kant places this "given" in the text, and the flow of his argument determines whether we will imagine *something* or *nothing*. Thus, what lies at the heart of the receptively given is a notion of otherness more functional than any imaginable character; it is the superlative functionality of an other whose very being or nothingness is determined by the imperatives of the subject's sovereignty. Without this other, Kant could not possibly retain the fantasy of unconditional sovereignty. Yet this oscillating otherness-not-otherness takes its own revenge as a persistent incomprehensible agitation in the heart of the transcendental narrative.[23] Nor have the pacemaking efforts of legions of Kant physician-scholars been able to steady its beat.

Perhaps we can now begin to see, exemplified on the screen of the most abstract epistemology, the general form of the dilemmas that are engendered by and that threaten an insidious version of (though not all) liberalism. This is a liberalism which seeks to establish its singularly supreme legitimacy on the basis of the vast array of diversity it claims to have respected and allowed into its order. If it truly allows this otherness to engage it, it will likely undergo transfigurations of differential and dialogical origin that will undermine its claims to singularly legitimate sovereignty. But if it denies such engagement, it will find its legitimacy and sovereignty convicted of arbitrarily betraying the very ideals on which it is to be esteemed. In the face of this "too much" or "too little," a radicalized liberalism might abandon the dream of sovereignty and willingly accept the risks of the former path. More often, however, an insidious type of liberalism seeks to establish its sovereignty in relation to the functionalized being and nothingness of its fictitious "otherness-not-otherness." The oscillating strategies of annihilation, exclusionary containment, and assimilation with respect to Native Americans provide but one very poignant case and point. The foundational narrative in the First Critique provides a monumental map of the contours of an imagination that so frequently engenders such problems. What are we to make of universalist respect in this context?

The Second Critique: Ironies of "I Must Not Receive"

Kant's moral theory as well as his reflections in the First Critique are animated and guided by the profound sense that disintegrative radical contingency accompanies receptivity. A problem analogous to the crisis of reception in theoretical epistemology sets the stage for Kant's transcendental moral narrative in the *Critique of Practical Reason*.

The dissatisfaction with the idea that in the most fundamental sense our knowledge of morality must be received externally (from a community, a tradition, positive law) can be gleaned in Francis Hutcheson's rejection of epistemological claims of scholastic Aristotelianism. Building upon some of the first Earl of Shaftesbury's arguments, Hutcheson claimed that the ultimate basis of our knowledge of morality was the "moral sense" which "receives" pleasing ideas (or, as Hutcheson later emphasized in his *Inquiry Concerning Moral Good and Evil*, "ideas of approbation") when confronted with benevolence.[24] Here, as Alasdair

MacIntyre argues, Hutcheson followed Descartes, Locke, and others in privileging "an epistemological stance embodying a first-person point of view. What-is is to be constructed from . . . what-is-immediately-present-to-me-as-idea."[25] With this move, Hutcheson drains moral reception of its radically external and imposing quality, resolving the most fundamental issues of moral epistemology and motivation internally by the nature of the moral sense. We "receive" simple moral ideas, but what we receive is given by our own nature, *in the first instance* unmediated by tradition, reason, social threats, revelation, and so forth. (On the basis of the moral sense, Hutcheson thought we could move deductively to a position that involved tradition, community, government, and scripture.) Beyond the inner resolution to questions of moral epistemology, however, for Hutcheson, we remain profoundly receivers. Thus the potentially radical subjectivism of his account—assailed by his critics—is checked because we receive our benevolent nature from God. This fundamental reception at the heart of human being is the ground on which the reception of moral heteronomy in other senses can be denied. Interestingly however (in a manner with certain resonances with Berkeley's understanding of God's being through the subject's immediate grasp of the "*esse* is *percipi*"), Hutcheson's argument for the divine excellence of God proceeds analogically on the basis of our own moral sense.

Hume, in his theory of the moral sentiments, intensifies the noncognitive aspects of, and makes more independent, Hutcheson's notion of moral sense. Claiming that God is at most an "is" from which no "ought" can be derived, Hume roots morality more resolutely in feeling itself, regardless of who made it or how. Finally it is the sentiments of pleasure and utility alone "in your own breast" that ground the conventions of justice. Morality cannot be received from another, whether that other be God (as the author of our moral sense) or (with the sheer exception of our very *sentiments* themselves) an order of nature.

For Kant, Hume illustrates that *all* moralities rooted in relations between ourselves and something else (e.g., objects, actions, ideas of perfection, God) are based on *feelings* we receive in these relationships, even where this basis is only tacit or is denied. These received feelings are fraught with the contingencies of human empirical sensibility, external facts, and their relations. Thus, they are incapable of yielding the a priori, necessary, and universal qualities that define moral experience for Kant. All attempts to ground morality in receptive relations to otherness are therefore as doomed as the analogous efforts to ground our experience of the objective world.

Kant's effort to solve problems of moral foundation traverses paths similar to those of his epistemological narrative. He must show that we give the law we cannot receive. In the case of moral reason, however, the story of our sovereignty does not immediately hinge on the precarious interplay between receptivity and spontaneity found in the *Critique of Pure Reason*. For the rational will does not legislate to a nature that is more than its *ex nihilo* creation but, rather, to itself alone. Hence the moral odyssey leads us initially into the realm of purely self-giving legislation. Curiously, though, receptivity becomes central later in the argument when Kant discusses the highest good, God, and immortality. Are these discussions mere lapses in an otherwise sound project? Or are they manifestations of the disruptive ironies, troubling paradoxes, and tragic blindnesses that plague the effort to be sovereign givers of the moral law?

I begin this discussion with a sketch of Kant's critique of the theories that ground morality in the will's relation to something other; for his claim that such theories disintegrate amid the contingencies of receptivity is central to his effort to establish autonomy as the "only" correct foundation for morality. Kant writes:

> All the confusions of philosophers concerning the supreme principle of morals [stem from the fact that these philosophers] sought an object of the will in order to make it into the material and foundation of the law (which [law] would then be not the directly determining ground of the will, but only by means of that object referred to the feeling of pleasure or displeasure). . . . [W]hether they placed this object of pleasure, which was to deliver the supreme concept of the good, in happiness, or in perfection, in moral feeling, or in the will of God—their fundamental principle was always heteronomy.[26]

According to Kant, when the will makes an object outside itself the determining ground of the practical rules governing desire, it ultimately surrenders itself to sensual pleasure, a relation of agreement between an object and desire.[27] Kant is not equating such heteronomy with crude hedonism. The object of one's pleasure may be the pleasure of another or the contemplation of God.[28] Yet, insofar as the idea of an

object is taken as the determining ground of the will, the will is ultimately attributable only to present or anticipated pleasure in the presence of this object. And when pleasure founds the *relation* between self
and other, it erodes both self and other.

Making pleasure foundational erodes the self because pleasure, "as a
receptivity belonging to inner sense" (58; 60), draws us into a sensual
relation with that which is outside the will. By grounding our aims on
such receptivity or susceptibility (*Empfänglichkeit*), we sacrifice our rational autonomy. We surrender ourselves to a susceptibility (an ineliminable but not determinative aspect of our will) which is contingent
on the presence of other objects and accidental aspects of the self (particular feelings) which may change uncontrollably. Kant's point is not
that we are determined in such cases by the object or by our own
natural being, but that we spontaneously legislate our own will according to the effects of sensual relations. The erosion of the self—its dis-
integration in the movement of dependent desire—is finally, in Kant's
view, the self's own doing.

Determining the maxims of one's will according to the agreeableness
of objects erodes the other in two related ways. The first is that our
practical relations with others and otherness are thereby governed according to "the general principle of self-love or one's own happiness"
(22; 20). The second is that, within this overarching self-love, there are
only questions of quantitative degrees of pleasure, questions of "how
much." No matter how diverse the objects of desire, they always effect
"the same life-force" and engender the same feeling of pleasure which
varies only in quantity (23; 21). Kant's focus on this reductionist characteristic is not meant to obliterate important distinctions. Thus, ego-
tistic self-love is far less preferable to him than a self-love that takes
pleasure in another's happiness—a desirable disposition. The point,
however, is that even the latter pleasure, taken as the fundamental determinant of our will, eclipses the moral distinctness of another rational
autonomous being as an absolute, unexchangeable end in itself. Self-
love provides an insufficient acknowledgment of personhood, which
deserves unconditional respect no matter what our natural inclinations
happen to be. Anything less is immoral.

Hence, moral principles based on receptivity are essentially entwined
with contingency and with reductions that nullify the very essence of
morality. For Augustine, of course, reception was not *fundamentally*
problematic, because facing the presence of God's infinite being was

synonymous with witnessing both his highest goodness and his charitable command that we receive him and obey his will. Our desire for him
was different in kind from desire aimed at any *object*, and receiving
God's love was at once qualitatively different from any concupiscible
pleasure and totally compelling. For Kant, however, the absolute moral
value of anything other than our will could not possibly emerge in
relation to that otherness itself because of the arbitrary and relative
aspect of all such relations.

While the first two theorems in the Second Critique identify these
errors, theorem three expresses Kant's first constructive step toward a
lawful moral imperative: *I must not receive*. Kant shifts to purely formal-
ist terrain in order to "abstract from" (27; 26) and be wholly "independent of" (29; 28) the destructive essence of all morality grounded in
material receptivity. This impulse clearly indicates a negative freedom.
Yet the will *not* to receive is simultaneously a positive, autonomous
freedom. By determining oneself so that "the maxim of one's will could
always hold at the same time as a principle establishing universal law,"
empty independence is given "positive definition" and "objective reality" (48; 49) as a causality that the will *gives itself*. Thus: "The sole
principle of morality consists in independence from all material of the
law (i.e., a desired object) and in the accompanying determination of
choice by the mere form of giving universal law. . . . That independence . . . is freedom in the negative sense, while this intrinsic legislation . . . is freedom in the positive sense" (33; 33).

Positive freedom requires that the will determine itself from universal law. Yet this requires in each case that we take the fundamental and
general injunction of pure practical reason and give it a *specific* rendering, a specific content or object, without which it would remain empty
and inoperative. To give unconditional law specificity means discerning
in each case whether the object of one's action is possible and necessary
for a morally governed will. As Kant summarizes: "A practical rule of
pure reason, as *practical*, concerns the existence of an object, and, as
practical *rule* of pure reason, implies necessity with reference to the
occurrence of an action" (68; 70). If pure reason could not give desire
an object, it could never be practical. To do so requires a "type" which
provides an example that allows us to consider the object, or "order of
things," implied by the universalization of the maxim of our action so
that we can judge whether or not the maxim is a specific instance of the
general rule. The typic facilitates judgment by providing an imagina-

tion of a world governed according to the maxim before which one asks "whether, if the action you propose should take place by a law of nature of which you yourself were a part, you could regard it as possible through your will" (69; 72). By thus imagining myself as an omnipotent lawgiver to a supersensuous nature of which I am at the same time a member, I can bit by bit establish moral sovereignty over myself. Without a typic, we "could make no use of the law of pure practical reason in applying it to that experience" (70; 72), and thus we could not establish our sovereignty.

In Kant's epistemology we become sovereign lawgivers to nature by categorically subjecting the manifold of intuition to the unity of apperception. Similarly, Kant's moral narrative concerns "the a priori subjection of the manifold of desires to the unity of consciousness of a practical reason commanding in the moral law, i.e., of a pure will" (67; 65). In contrast to the "combining" in the First Critique, however, the "subjecting" in the Second Critique is such that the moral law "limits its material, must be a condition for adding this material to the will" (35; 34). Since nature, broadly construed, "is the existence of things under laws," the subjection of the manifold of desires forms a "supersensuous nature" (43; 44) in which, in contrast to sensuous nature where objects are "the causes of conceptions which determine the will," "the causality of objects has its determining ground solely in the pure faculty of reason" (44; 46).

For Kant, the subject appears to have established a sovereignty of such lucid power that it dissipates all opacity concerning "what is required of us." Indeed, "even the commonest and most unpracticed understanding without any worldly prudence" can see what must be done, "easily and without hesitation" (36; 38). Having rejected murky considerations of happiness, physical capacity, the contingencies of myriad conflicting physical causes, finitude, and so forth, one considers the moral questions before one simply from the self-given perspective of the mere form of law.

One wonders, though, whether this sovereignty is achieved only by means of an arbitrary and unacknowledged conditioning—by an internalization and reification of radically contingent limits. How sovereign, reasonable, and universal is the studied myopia of a being that must endlessly reject concern for the material finitude (of perception, judgment, desire, relations with others) that is an ineliminable part of its being in the world? Is not such sovereignty gained only at the price of self-imposed exile from the terrain of one's own and others' very being? For what type of very limited reason would this choice constitute an incontestable "fact"? And if pure practical reason is determined in oblivion to the contingency of its limits, must not this oblivion mark the object of its determination? Does not such a reason, so constrained in its object, appear radically conditioned, particular, impotent, limited, even irrational? Would not a "necessity" that could not dispel radical self-doubt when faced with its own object begin to tremble?

Kant formulates extensive responses to these questions in the "Dialectic of Pure Practical Reason," though few have found his efforts satisfactory. Yet perhaps it is to Kant's credit that he does not try to ignore these questions and pursues the narrative of receptionless lawgiving sovereignty with such tenacious integrity. Kant's tenacity pulls him down a path of reception every bit as radical (and implausible) as the "I must not receive" which it is intended to supplement and secure. Still, this strange and circuitous journey illuminates far more than many of the supposedly straight and rational routes chosen by Kant's followers, who thought that one could abolish Kant's problems simply by lowering the volume on his aspirations. Tracing Kant's circuitous journey, we turn to his discussion of the highest good.

Though considerations of happiness must not be the supreme determinants of our will, Kant nevertheless writes: "Man is a being of needs, so far as he belongs to the world of sense, and to this extent his reason has an inescapable responsibility from the side of his sensuous nature to attend to its interest and to form practical maxims with a view to the happiness of this and, where possible, of a future life" (63; 61). Kant's claim here is not simply that the body has needs which must be attended to for survival and happiness; nor is it—as he makes clear elsewhere—that pure practical reason has an indirect duty to ensure a basic level of happiness for the sensuous self because miserableness and resentment are conditions that severely distract us from our moral duties.[29] Beyond these, or subsuming them, the claim is that *reason itself* "has an inescapable responsibility," a responsibility rooted in its essence as unconditional autonomy in the context of finite beings, to legislate "with a view to . . . happiness." How can this be?

For Kant, rejecting happiness as the supreme determinant of our will is utterly different from rejecting happiness as a vital concern. In fact, rejecting happiness as the supreme ground *and* making it a vital concern are *both* necessary for the truly autonomous will of a "finite rational being." For if rejection were reason's sole relation to sensuous inclination and pleasure, reason would reveal itself as simply another finite, conditioned, particular, contingent kind of thing: an activity merely exclusive of and negatively defined in relation to the array of contingent things which it is not. Kant confronts this problem when examining the unconditional object that articulates the unconditional will, based on the injunction not to receive. Paradoxically, this *articulation*—which gives unconditionality its content, definition, and *object*—requires a thoroughgoing reception. The happiness that is repressed in the establishment of the fundamental law of practical reason returns in grandiose fashion as reason defines its absolute unconditionality in an object wherein happiness abounds, though within the limits and according to the proportions prescribed by the moral law itself. Unconditional positive freedom, in order to discover its unconditionality (instead of its limits) in its object, must give itself the reception it had to deny in producing itself.

Moral law, and virtue as one's endless progress toward alignment with that law, is the "supreme condition of whatever appears to us to be desirable" (110; 114). However, it is only a part (albeit the unconditional condition) of the highest good, not the whole. The highest good also includes "happiness in exact proportion to morality" (110; 115). Kant writes:

> Happiness is also required [as part of the highest good], and indeed not merely in the partial eyes of a person who makes himself his end but even in the judgement of an impartial reason, which impartially regards persons in the world as ends-in-themselves. For to be in need of happiness and also worthy of it and yet not to partake of it could not be in accordance with the complete volition of an omnipotent rational being, if we assume such only for the sake of the argument. (110; 114–15)

A being of impartial reason, argues Kant, regards itself and others as ends in themselves[30] who deserve (and for whom we must will) happiness in proportion to morality. Not to will this deserved happiness would be to flee in fear of a contamination of unconditioned reason by the self's weak and conditioned being in the phenomenal world (where one's own and others' happiness is so elusive) and irrationally to fabricate an unconditionality claiming no fundamental relation to happiness and its problems. Yet the conditionedness of such an unconditionality appears when we imagine an omnipotent rational being. Just as imagining ourselves as sovereign lawgiving members of a supersensible nature in the typic was to illuminate partiality and impartiality concerning questions of law and receptivity, imagining an omnipotent rational being allows us to imagine an object of reason unencumbered by limits of power and judgment. In this light, a being who does not will happiness in the highest good is partial, conditioned, not fully rational. This partiality contrasts with the true impartiality of an omnipotent rational being, not by placing inclinations first but by incorporating phenomenal finitude into the understanding of reason's essence. By denying happiness in its highest object, this partiality makes conditioned finitude determinative of reason (now partially blind and limited). Yet how could this conditioned object stand for a being whose moral will lies precisely in its unconditionality? Must not the unconditional will, born in "I must not receive," inevitably lead to an unconditioned object that includes happiness in accordance with moral worth if it is not to flounder in contradictions? And since such a will cannot give itself the happiness that it is obligated to desire, must it not somehow *receive* that happiness? "I must not receive" seems to become entangled in "I must receive."

Analogous to the way moral law requires the practical judgment of objects as types in order to gain in each situation determination and meaning, at the deepest level of Kant's moral project, the fundamental law of pure practical reason *as such* remains empty without *its* object. But what is the endpoint and totality which would express the absolute unconditionality of the unconditioned itself? Kant responds: "Pure practical reason . . . seeks the unconditioned . . . and this unconditioned is not only sought as the determining ground of the will but, even when this is given (in moral law), is also sought as the unconditional totality of the object of pure practical reason . . . the highest good" (108; 112). The unconditioned determining ground of the will and the unconditioned totality of its object *must* be sought together, not only because the former is the ground of the latter but because the latter gives meaning to the former. The highest good as total object responds to the question "What is the highest realization of moral law and, therefore, the ultimate object of our will?" Without an ultimate object, *this* object, the idea of the totally unconditioned reason is no more than an utterly vacuous intransitive tautology. Hence, Kant writes that "the furthering of the highest good . . . is an a priori necessary object of our will and is inseparably related to the moral law" (114;

118). If reason did not take the highest good as its object, its unconditionality would either remain empty or forfeit its essence as unconditioned and impartial (and thereby forfeit its absolute privilege over all heteronomy). Yet if pure practical reason must have an object, and if it must connect virtue and happiness, a question immediately emerges concerning the possibility of such a synthesis. For moral law commands "through motives wholly independent of nature and its harmony with our faculty of desire" (128; 125), while willful achievement of happiness is "dependent not on the moral intentions of the will but on knowledge of natural laws and the physical capacity of using them." From this situation, Kant provisionally concludes that we can expect "no necessary connection" between virtue and happiness (113; 118).

If, however, the moral law commands pursuit of the highest good, and if the latter contains virtue and happiness in *necessary* connection, and if no such necessity is thinkable, then moral law, far from exemplifying universal rationality, would involve reason in the contradiction of commanding that it determine and will itself according to something which is fundamentally impossible. The unconditioned object of an unconditioned will would be null because the conditions in which it would cohere with necessity are inconceivable for us. The unconditioned will, by the impossibility of its necessary object, would reveal itself as illusory. Thus Kant writes that "the impossibility of the highest good must prove the falsity of the moral law also. If . . . the highest good is impossible according to practical rules, then moral law . . . must be fantastic, directed to empty imaginary ends, and consequently [be] inherently false" (114; 118). Suspense mounts. The leading character in the narrative of self-given moral sovereignty appears to be on the verge of a breakdown. Only a God can save us now.

Reason postulates the existence of "God as necessarily belonging to the possibility of the highest good" (124; 129). God as the author of nature is the "ground of the exact coincidence" between happiness and reason (125; 129). God is not the heteronomous commander of our will, nor a provider of heteronomous incentives as in the Canon in the *Critique of Pure Reason* (A813 and 818, B841 and 846).[31] Instead, God is rooted in the unconditional moral law itself. God is a giving other that autonomous reason gives itself—postulates—through a combination of duty ("the endeavor to produce and further the highest good in the world") (126; 130) and a subjective need to conceive of the possibility of the object of this duty.

Hence receptivity, the rejection of which was central in Kant's moral narrative, returns here in grand form. What begins as an unconditional rejection of receptivity-as-contingent-desire ends in the need to postulate an ultimate harmony between our self-giving moral autonomy and the pleasurable reception of the heteronomous world, a harmony in which we pleasurably receive heteronomy thoroughly in proportion to our ability to give ourselves moral autonomy. Without such harmony, the unconditioned will would be devoid of the unconditioned object— and thus would be empty, imaginary, and ultimately lacking a priori sovereignty. To overcome heteronomy, the pure will must now, according to its own law, postulate the other (nature and its author) not as threateningly contingent but as the morally generous condition of the object of our autonomy and, therefore, ultimately, autonomy itself. Self-giving sovereignty must give itself generous otherness to sustain its unconditional rejection of the contingencies involved in receiving heteronomy.

This path of thought appears in the subject's relation to time as well. "Pleasure," temporally defined, is "consciousness of a presentation's causality directed at the subject's state so as to *keep* him in that state. [D]ispleasure . . . contains the basis that determines the subject to change the state."[32] Pleasure is consciousness of a relationship to something present such that the subject is drawn to extend this relation in time. Thus, when Kant enjoins us to reject pleasure (receptivity) as the ground of our moral will, he is, temporally speaking, enjoining us to cease making determinations based on particular modalities of time-consciousness (yearning for extension, alteration, or cessation) entwined with given presentations. Indeed, as exemplified by the moral rightness of (a) accepting death from a malicious sovereign instead of making a false deposition and (b) rejecting suicidal inclinations and thoughts, the true character of morality appears in the rejection of contingent, emotional time-consciousness as will-determining—not simply in relation to particular presentations, but in relation to the termination or extension of one's life as such (i.e., the sum of one's relations to the world and time).

Yet, in refusing to *receive* as foundational one's particular and general relations in time, the moral self must postulate—or *give itself*—the idea of the immortality of its own (and the other's) soul, the "infinitely enduring existence and personality of the same rational being." For Kant, this postulation of the soul's infinite duration is "an inseparable corollary of an a priori unconditionally valid practical law" (122; 127). The moral law which is based on "I must not receive" emotional time-concerns must, finally, receive an infinite time—in the form of a gift it gives itself—in order to articulate itself.

We have seen that the unconditioned law must will an unconditioned object, and that "complete fitness of intentions to the moral law" is the highest good's supreme condition (122; 126). Willing less than complete fitness would mark the will as conditioned (e.g., having internalized temporal limitations, temporally borne contingencies, internal weakness, and so forth). Yet Kant maintains that, for a rational being, "complete fitness . . . is holiness [having no desire at odds with moral law], which is a perfection of which no rational being in the world of sense is at any time capable" (122; 126). As receptive beings, our desire is simply too manifold and opaque for us ever to achieve complete accord with the moral law in every fiber of our being. Thus it would appear that this aspect of the object of the moral law (and with it the moral law itself) would be impossible, fantastic, empty. Unless.

Unless we postulate the idea of an immortality that is the time of an endless moral progress which God views as a whole to be conformable to morality. Without receiving this infinity of time, our sovereignty would come up short: its object, as impossible, would be irrational. With infinite time we have a realm of striving that never ends, but as never-ending it is never definitely limited, closed, or impossible. Thus, in order to reject time's unfriendly invasions of our autonomy, we require a moral law that must give itself the postulate of a friendly (i.e., infinite) time; we require the unending possibility of moral improvement wherein resides the intelligibility of the will's unconditional object.

These arguments for God and immortality as necessary for moral law have a dissatisfying ring. And not simply because, as Hegel and Nietzsche were aware, God's being chokes on the breath of human sovereignty (let alone the idea that God somehow "is" *only as a practical postulate of that sovereignty*). More centrally, self-given moral autonomy, which Kant grounded on a rejection of foundational receptive relations to otherness, *itself seems to choke* on his assertion that human sovereignty is empty and false unless one postulates an ultimate reception of this otherness (albeit now drained of recalcitrance: generously authored nature, infinite time) as the condition of moral law's unconditioned object. Kant seems to be arguing *against* the possibility of an uncondi-

tional autonomy that draws its sustenance exclusively from rational beings with very short lives, limited power, lots of blindness, permanently
recalcitrant desire, and little faith in a God who will make up for this
finitude. Kant thus seems—especially to his more secular followers—to
splash water on the lambent flame of sovereign rational autonomy he
has so carefully tended (a sovereignty confident of itself and ideally
unmoved by the finitude that washes about it, a reason that would cling
resolutely to itself and without myth even in the face of its impotence
in the world). Yet these twists and turns signify Kant's rigor, not his
weakness, in thinking through the possibilities and difficulties of the
rational subject's moral sovereignty.[33]

The weakness of this narrative of the sovereign subject for establishing unconditional morality is one reason why contemporary theorists
like John Rawls and Jürgen Habermas have sought different stories.
We shall explore Habermas's approach in Chapter 3 below. For the
moment, however, I offer two examples which illustrate some serious
practical difficulties with the sovereignty narrative.

The Murderer at the Door and the Sweet Sense of Having Done Right

Kant's discussion of lying and revolution illustrates the deleterious
implications of his narrative of sovereign autonomy for morality and
politics. Contrary to his intentions, these examples illustrate not the
clarity and simplicity of moral imperatives but, rather, the unreceptive
hazardous oblivion that emerges from his fixation with pure self-giving.
These problems further illuminate the need to consider alternative narratives from which to draw an ethical position.

It has been noted that outside a context of interpretation, the meaning, limits, and definition of Kant's categorical imperative are difficult if
not impossible to understand.[34] An acute sense of this problem can lead
one to reject the empty formalism of Kant's morality or to elaborate a
way in which judgment mediates between duty and historical context to
give duty specific sense.[35] Yet this question of duty and *external* context
is not what most interests me here. Instead, I am concerned not about
the absence, but the *presence* of a context that frames and pressures the
manner in which universal morality is interpreted. This context is *internal* to duty, for it is precisely in *the narrative of sovereignty itself* that

duty is housed, explained, and philosophically justified. This foundational narrative dramatically influences Kant's interpretation of how the categorical imperative engages the phenomenal world.

It is commonplace in the secondary literature to disassociate Kant's moral principles from his illustrations of their application.[36] My claim is not that Kant's universalist morality is inseparable from his comments on, say, lying and revolution. Still, the insistence on self-given sovereignty in the Kantian narrative exerts a powerful pressure in the direction exemplified by Kant's illustrations. To ignore this fact is to overlook how the story which justifies a moral position frames the meaning and applications of that position; in concealing this important relation, one risks losing both a critical perspective on Kant and an insight into the functions of one's own "justifying" stories.[37] Kant's story of morality as sovereign self-given law leads him toward interpretations of that law which manifest the "clarity" and "certainty" appropriate to and confirming of a self-transparent reason. Complications, ambiguities, or uncertainties in his moral determinations, if Kant were to admit to them, would raise questions of difficult receptions of the phenomenal world that would belie and erode the narrative of moral sovereignty. A self whose reason is frequently torn and uncertain in the world would seem to enact a story very much at odds with Kant's foundations. Difficult applications might make the project tremble in ways Kant could never accept.

Hence, while Kant does, *very rarely*, write that "these laws require a power of judgement sharpened by experience," in order to decide where they apply and to "gain access to man's will and an impetus to their practice,"[38] comments like the following are *far more common* and constrain Kant's development of his examples: "[The] most unpracticed understanding without any worldly prudence" can "satisfy the commands of the categorical command of morality,"[39] and "I do not need any penetrating acuteness in order to discern what I have to do in order that my volition may be morally good."[40] Compelled by his sovereignty narrative toward illustrations that confirm the moral possibilities of the "unpracticed understanding," Kant's discussion of lying when in need of money must exemplify a simplicity that lucidly illustrates the unconditional. Frame your maxim: "When I believe myself to be in need of money, I will borrow money and promise to repay it, although I know I shall never do so." Ask if this maxim could be a universal law. Realize immediately that such universalization would render promises meaningless and impossible. Reject this maxim which "must necessarily contradict itself."[41] Though Kant mistakenly clouds this section with references to self-love, as Wolff notes, the real issue is to distinguish morality from prudence.[42] Clearly this test rules out both unselfish false promising (in order to buy necessary medicine for the dying grandmother) as well as selfish false promising.

Years later, Kant passes the most extreme test of his cursed lucidity and holds fast to his rejection of lying even "to a murderer who asked whether our friend who is pursued by him had taken refuge in our house."[43] If we lie, we "harm mankind generally, for [a lie] vitiates the *source of the law itself*" (347, my emphasis), because it both makes contracts impossible and violates the universal form of law. Truthfulness, within the framework of unconditional autonomy, is "an unconditional duty which holds in all circumstances." But is this not a self-giving autonomy which, as the "source of law itself," takes its own preservation and its foundational rejection of messy receptivity to be of incomparable importance and thus retreats into a most immediate and simplistic formalism—no matter what—in order to avoid entanglement in the difficult relations so erosive of sovereignty? Kant appears to sense that acknowledging ambiguity in *applying* universalist reason might undermine, or at least make questionable, the "I shall not receive" which grounds his reason. Precarious as it is, Kant's sovereignty narrative generates an insistent simplification and certitude that conceal its fissures.

So, Kant says, be truthful to the murderer at the door. Moreover, we do "not do harm to him who suffers as a consequence; accident [the contingency of the murderer's presence] causes this harm. One is not free to choose in such a case, since truthfulness (if he must speak) is an unconditional duty" (349). "Formally violating the principle of right" is much worse than the particular injustice, because the latter was not made a principle of one's will (350). Presumably, our truthful "friend" is, to quote from *Perpetual Peace*, "able to enjoy the sweet sense of having done right,"[44] for the concealment of one's moral damages is as integral to self-giving autonomy as is the concealment of this narrative's precariousness.

I disagree with Beck's assertion that Kant's "argument is very un-Kantian."[45] Yet who could disagree with Beck's claim that one could also construct a universalist argument for lying to the murderer? Couldn't we, with H. B. Acton, construe a categorical imperative which "does not entail that basic moral rules cannot have exceptions, but only that the permissible exceptions are universalizable maxims"?[46]

It is not obvious, however, that Acton's interpretation is the "correct" one. His rendering is more or less plausible depending on the "song of ourselves" by which it is justified, supported, elaborated, and oriented. My purpose here is not to explore neo-Kantian songs, but to note the effects of *Kant's* song, sung in the single key of unconditioned autonomous sovereignty. One of Kant's virtues is to be very clear that this song starts to go very flat as soon as one introduces conditions in one's improvisations, whether at the metalevel (e.g., rejecting happiness in the highest good) or at the level of microapplication. Kant knows that it is in the nature of his composition to self-destruct with the slightest modulation. Hence, "unconditional" must be strictly and literally interpreted if autonomy is to maintain the absolutely originary status through which it is sovereign. For Kant, a categorical imperative "with exceptions" is no longer categorical: far from self-given autonomy, it signifies a reception of the world's difficult discordance. What exactly is meant by a universal that allows exceptions on the basis of other universals? The word "exactly" is crucial for Kant: without it, his song never resolves. If there are many conflicting universals, how is our choice among them to avoid a dimension of the arbitrary? We ought to consider consequences, says Paton's Kant, but "in many cases remoter consequences ought to be ignored."[47] "Many"? "Remoter"? Are these the clear notes of unconditional sovereignty, or the muddled tones of a more vague, approximate, relative, hypothetical tune? And whence the sovereignty of a reason engaged in the difficult negotiations of finite receptivity? Kant's answer is simple and appropriate to the narrative through which the world acquires meaning for him: *Never* lie.

Kant treats rebellion in like manner. His rejection of revolutions based on a regime's illegitimate origin or relation to suffering is not surprising, since the questions of right and legitimacy are no more empirical than the moral law to which politics ought always "bend its knee." A state's legitimate authority is based solely on the regulative idea of the supersensible original contract that "arises from [each person's] own lawgiving will" (316; 127). But what if the sovereign clearly, systematically, and repeatedly violates the "universal principle of right" which states that "any action is *right* if it can coexist with everyone's freedom in accordance with a universal law, or if on its maxim the freedom of choice of each can coexist with everyone's freedom in ac-

cordance with universal law"? (230; 56). Can a revolution ever be ac-
ceptable under these conditions?[48]

No, never. As in the case of lying, numerous interpreters question
whether Kant's answer follows consistently from his universal principle
of right and original contract.[49] But I think Kant is consistent here, and
though there are clearly other interpretations with substantially different outcomes which are also consistent with universalist morality and
right, Kant's understanding of sovereignty (operating on two levels this
time) drives him in the direction with which we are familiar.

A people must not lay claim to a right to revolt when their rights are
injured, for by establishing such a right they would be willing a contradiction. They would, in the name of right, simultaneously be willing
both a ruler (as a necessary condition of the possibility of right) and the
possible rule by the ruled over the ruler—an absurdity that can hardly
be thought of as a universal law capable of withstanding public expo-
sure.[50] Kant raises a genuine problem here, and it is one that becomes
especially difficult as one moves closer to sovereign formalism. "For
someone who is to limit the authority must ultimately have even more
power than he whom he limits. . . . This is self-contradictory, and the
contradiction is evident as soon as one asks who is to be the judge in
this dispute between people and sovereign" (319–20; 130–31). By re-
belling, the people are inconsistent with right because the maxim of
their action annuls the possibility of sovereign rule, while "a rightful
condition is possible only by submission to its general legislative will"
(320; 131).

In part this extreme formalism, characterizing both Kant's posing of
the problem and his solution, can be accounted for in terms analogous
to my analysis of Kant's position on lying.[51] Kant's foundational narrative pressures him toward modes of applying autonomous reason that
exemplify logical transparency, simplicity, and certitude—for, otherwise, he would be drawn into difficult efforts to interpret a messy world
in ways that might raise questions about sovereign unconditional reason. Yet Kant's single-minded effort to avoid immediate self-contradic-
tion here is overdetermined. For the starkness of the contradiction can
itself be comprehended only in conjunction with the way in which the
sovereignty narrative functions in his "rational Idea" of the condition
of people with and without public law. According to Kant, prior to the
establishment of "public lawful external coercion," humans exist in a
state profoundly analogous to Kant's discussions of the manifold in the

first two Critiques. Our condition is one of indeterminacy and incoherence, devoid of any durable relations, and is self-destructive (as in the Second Critique). Just as the manifold alone in the First Critique is unworldly in terms of experience, and that of the Second Critique lacks worldness in terms of morality, the manifold of the state of nature is civilly worldless: "wild," "lawless," "never secure," "*status iustia vacuus*" (316, 312; 127, 124).

The worldness of the civil must be determined by public law. Hence, each lawgiving being must realize that "the first thing it has to resolve upon is the principle that it must leave the state of nature . . . unite itself with all others, [and] subject itself to public lawful external coercion" (312; 124). For Kant, the ideal civil condition is a representative republican government. Yet even a despotic ruler gives law and unity in the fundamental sense that establishes a civil world, no matter how flawed. Any ruler brings forth an empowered image of the social contract and in this sense plays a role somewhat analogous to the schema (the "universal procedure of imagination in providing an image for a concept") in the *Critique of Pure Reason* (B179–80, A140).[52] To actively resist a sovereign is thus equivalent to willing against a faculty that establishes civil being. It is to will nothingness: *status iustia vacuus*.

The starkness of the contradiction brought on by willful resistance, then, is rooted in the narrative of sovereignty that constitutes Kant's account of the will and the civil. That account is based not on experience (Hobbes's claim) but on an a priori rational Idea (312; 124).[53] But this perhaps is only to say that Kant's account is produced by an imagination enthralled by a strange tale. Other accounts of our condition might illuminate other possibilities.

So perhaps it is time to explore some alternatives. For have we not seen repeatedly in Kant the untenability of this most powerful and rigorous effort to construe human subjectivity as the self-giving ground of truth, morality, and the civil? Have we not seen, in the First Critique, this ground's need to receive an "other-not-other" that oscillates incoherently between being and nothingness according to the demands of a transcendental narrative through which the "law-giver of nature" might be able to imagine and recognize itself? Have we not seen, in the Second Critique, the need of the unreceiving giver of the moral law to receive the generous alterity of an infinite time and a benevolent God—an alterity, oddly enough, which it gives itself in the postulates? And have we not seen, in the context of the troubling insistences of this sovereignty, a practical orientation that tends to eclipse the other—even as it proclaims to give universal respect? How likely is it that such a sovereign self will give a respect to others that will not self-destruct in the insistent blindness of the "I must not receive"? It will be said that the autonomous lawgiver is bound to receive the other in the injunction to treat others as ends. But the meaning of this injunction, the meaning of human subjects as "ends," is so overdetermined and obfuscated by the requirements of the sovereignty narrative that the other-as-a-question, as an elusive ambiguous specificity the reception of which would be integral to the opening of so many difficult ethical relations, is persistently concealed. In this concealment, we are continuously locating the other for the "murderer at the door" while simultaneously enjoying our "sweet sense of having done right." Of course, it would be crazy to reduce Kantian respect to murder: we are too indebted to it, for all its severe problems, to embrace such a brazen ingratitude. Still, in every Kantian lawgiving there is a little or a lot more "murder," more blindness and violence, than is necessary even for beings so finite and prone to eclipsing otherness as we are.[54] In search of less violence and a grander generosity, I turn to other magnificent corners of Kant's corpus.

Genius and Aesthetic Ideas: Gestures beyond Sovereignty

I have traced some of the ways that the transcendental story of sovereignty plays itself out in Kant's epistemological and moral writings. Kant formulates this narrative in an effort to extricate humans from the seemingly insurmountable difficulties we face when we imagine ourselves as fundamentally receptive to otherness in our being, knowledge, and moral stance in the world. Yet, as we have seen, the story of sovereignty has serious difficulties.

Faced with these problems, we must search for other understandings of ourselves that might hold more power to illuminate the human condition and broadly solicit our actions in more desirable directions. I contend that there are evocative currents in Kant's writings that radically contest his central narrative.[55] These currents are frequently twisted back, by Kant and many of his interpreters, into the mainstream from which they threaten to break away, like an eddy that laps the banks only to be sucked back into the central flow.[56] What follows, then, is perhaps less where Kant actually went in the Third Critique than it is where he *might* have gone had he "missed the curve." The journey beyond sovereignty is not without its dangers; but we proceed anyway, thrown by Kant's unwitting illuminations of the impossibility of traveling the established channel any longer.

In the rich sections in the *Critique of Judgment* concerning genius and fine art, as well as in certain passages on the sublime, there are suggestions of an alternative terrain for understanding selves in relation to the world around them. It is possible to read these sections as further articulations of solar self-given subjectivity, consonant with the loudest and most persistent voices in Kant's discussion of the beautiful and the sublime. Concerning the beautiful, a judgment of taste has its determining basis in the subjective mental state wherein our powers of imagination and understanding, when prompted by a given presentation, are brought into a harmonious free play that "quickens" these powers and gives rise to a feeling of pleasure. In judgments of taste, the subject passes beyond determinate lawgiving to a freedom unconstrained by determinate concepts yet nevertheless sovereign insofar as the judgment it gives from within has universality and necessity; the subject, moreover, demands that others recognize this judgment as universal. There is nothing profoundly relational between the subject and the world in judgments of taste: beautiful objects simply trigger the free play of the cognitive powers within the subject. Still, insofar as the object prompts us thus, we attribute to it a formal purposiveness with respect to our powers, and we thus recognize it as nature's confirmation of our sovereignty. Concerning the sublime, Kant's dominant argument leads to a relation between imagination and reason that exalts the sovereignty of our self-given reason beyond anything nature is capable of presenting.

However, though it is *possible* to read Kant's sections on genius and aesthetic ideas as consonant with these themes of sovereign subjectivity, such a path is not very fruitful philosophically nor very reflective of the direction of thought that those sections most express.[57] I approach an alternative interpretation in the context of Kant's more marginal, yet most profound, reflections on the sublime; for they introduce perceptual and cognitive themes essential for exploring the later sections of the *Critique of Judgment*. My discussion of the "marginal sublime" begins with a brief overview of the "sovereign sublime." This contrast highlights the epistemological breakdown of sovereignty and the point at which Kant suggests a very different understanding of the genesis of self and world.

The sovereign sublime occurs when an object prompts us to present the unboundable. Taking the mathematical sublime as paradigmatic, we experience the sublime when the imagination, *apprehending* partial presentation after partial presentation of a large object, finds itself incapable of *comprehending*—or collecting and holding together—the parts into a bounded unity. The object exceeds the imagination's comprehensive capacity insofar as the "presentations . . . that were first apprehended are already beginning to be extinguished in the imagination, as it proceeds to apprehend further ones" (252; 108). The imagination's inadequacy before such immense magnitude (or, analogously, power) is a source of displeasure. Yet Kant rejects Edmund Burke's interpretation of this experience as a terrifying awe in the face of overwhelming otherness that leads to a humbling sense of finitude.[58] Indeed, Burke erroneously substitutes respect for the object of nature for what is truly deserving of respect here: namely, our rational subjectivity. For Kant, the object merely "makes intuitable . . . the superiority of the rational vocation of our cognitive power over sensibility" (257; 114).

This notion becomes clear when we trace the hegemony of reason in Kant's narrative. First, the imagination's effort to comprehend an appearance as a bounded whole is not presented as intrinsic to its very being, but rather as "an idea enjoined on us by a law of reason" (257; 114) which seeks absolute totality. Hence, imagination's incapacity to accomplish bounded closure is not a failure before an overwhelming manifold to generate a sovereignly legislated world (suggestive of a deeper and stickier finitude), but merely a failure to satisfy reason's demand for a presentation of the absolute. This failure instills in us a great respect for, and pleasure in, our subjectivity as we recognize that even the faculty of sensibility is governed by a rational vocation that exceeds sensual nature. Furthermore, as our imagination fails to combine the unbindable, as it failingly strives toward what is for it infinity, it "arouses and calls to mind" ideas of reason—truly infinite totality—which, it realizes, absolutely exceed the large yet ultimately finite objects of nature. We recognize that nature is small in comparison with our reason and, thus, we recognize our superiority.

Yet as one explores Kant's quite variegated account of our encounter with nature's "boundlessness," "chaos," "wildest and most ruleless disarray and devastation" (246; 99–100), one senses that he includes and evokes more than the limits of the sovereign sublime can contain. One senses that subject-given reason ought not remain quite so unscathed and exalted—that Kant's efforts to comprehend our encounter with the boundless manifest a certain concealment which is emblematic of the concealments that reason itself must enact in every experience of the Kantian sublime. Hence we turn to the project of bringing forth from the margins the strange experience of the boundless and its centrality in the genesis of the self and its experience of the world.

Kant asserts that the imagination's effort to comprehend the object is owing to an idea of reason, suggesting that the failure of this effort stems from the excess of the subject's guiding rational sovereignty, rather than from an inherent failure of the imagination before otherness which would pose a fundamental challenge to this sovereignty. Yet, from Kant's account of the synthetic imagination in the *Critique of Pure Reason*, it is clear that the imagination's collecting together into one is intrinsic to its pivotal role in the genesis of experience.[59] As we have seen in the first-edition Transcendental Deduction, experience is born when "unity of intuition arises out of the manifold" as the latter is "run through and held together," requiring both a "synthesis of reproduction in imagination," such that each of the parts is reproduced as one advances to the next, and a "synthesis of recognition in a concept" which establishes an "identity [unity of the object] in the manifoldness of its representation" according to a priori rules.

The central point here is that at the heart of the possibility of *all experience* is a (preconscious) striving by the imagination and understanding to grasp the manifold and hold it together into one. When this effort is exposed to failure, our very experience of the world begins to tremble and come undone. And we know that insofar as the world unravels in this manner, so too does the very unity and being of the sovereign subject; for subjectivity, as the unity of apperception, is possible only under a synthetic "unity through which all the manifold given in an intuition is united in a concept" (B139; 157). Lacking this unity, the sovereign self would be lost and disintegrated in the "crowding in" and "blind play" of the manifold. There would be no necessary continuous relation from one presentation to the next such that they would all be "mine." Without a comprehended world, the sovereign subject *is not*. Hence, the experience of the sublime does not stem most primordially from a sovereign reason that would generate and remain immune to a failure to achieve a magnificent unity of apperception; rather, it is intrinsically linked to the comprehending movement at the heart of the genesis of self and world. The experience of the sublime illuminates a failure to accomplish the bounded closure required for sovereignty. And the experience of this failure is sharply painful because the aroused sense of nonsovereign boundlessness not only refers to the present moment of the sublime encounter but also illuminates a ubiquitous and profound contingency, boundlessness, and even violence that is inextricably entangled with the genesis of both the self and *all* worldly experience. For this part of the story, passages in the Third Critique are far more helpful than the more sanitized sovereignty of the Transcendental Deduction in the First Critique.

In contrast to the latter, the Third Critique's margins tell a story of the comprehending activity that collects the manifold of inner temporal sense and holds it together into a unity which is wrought with a distinctly tragic sensibility:

> Comprehending a multiplicity into a unity (of intuition rather than thought), and hence comprehending in one instant what is apprehended successively, is a regression that in turn cancels [or "abolishes," *aufhebt*] the condition of time in the imagination's progression and makes *simultaneity* intuitable. Hence, (since temporal succession is a condition of the inner sense and of an intuition) it is a subjective movement of the imagination by which it does violence to the inner sense, and this violence must be the more significant the larger the quantum is that the imagination comprehends into one intuition. (258–59; 116)[60]

Kant alludes here to a violence at the heart of the comprehending which is necessary for the intuition of simultaneity, or spatiality, of the world in which the self coheres. He roots this violence in a cancellation of the manifold's as yet unworldly, unbounded, untamed succession. And this violent aspect becomes "more significant" as the quantity of unruly/unruled time which is to be bound into one increases.

But how are we to understand this violence? If we follow Kant's account in the First Critique, concerning the manifold of intuition as an other-not-other that oscillates between being and nothingness in order to satisfy the requirements of the sovereign subject's legislative synthesis, it is difficult to imagine what could be violent about holding together the manifold into a unity. Yet if we follow the most provocative moments of the Transcendental Aesthetic, we can discern the beginnings of an account of the self's encounter with otherness that is prior to its more active syntheses, an encounter where multiplicitous presentations are given, or begin to be received, but are not possessed; are given with an indeterminacy which is neither being nor nothingness, which is pregnant with possibility but recalcitrant by virtue of this same inexhaustibility, which is partially formed through a "synopsis" that precedes more active and determinate "syntheses." From this perspective, we can begin to see a relation between comprehension and the temporal movement of inner sense that, while expressing a possibility of the latter, simultaneously transgresses its protean, partially formed, yet unbounded character with a certain violence. In "cancel[ing] the condition of time in the imagination's progression" in order to bring forth simultaneity (a world of unified objects more determinately there) synthetic "regression" is simultaneously a *reduction*. In gathering together into one, comprehension reduces the manifoldness and possibility of the indeterminately given that is sensed in the apprehension of inner sense; multiplicity is reduced to accommodate the conditions of unity. The essentially wild and pregnant presentations of the manifold are "run through" in light of (and subordinated to) the requirements of a determinate ordering in which they "hold together" in a recognizable object. This light may well express and bring to a certain fruition and magnification latent possibilities (of identity and difference) of the manifold (and in this sense perhaps *aufhebt*, in the passage just cited at length, also means "picks up and preserves"), but it also reduces and eclipses possibilities whose expression might involve the unraveling of a given order and unity. In temporal terms, the extended progression of inner sense signifies (even if it does not achieve) a duration in which this multiplicity and pregnancy of the indeterminately given might be sensed. The violent aspect of the reduction of the manifold can be articulated not only in relation to "internal" richness and possibility but also in relation to the object and its "exterior"; for qualities and possibilities of relation between manifold presentations are (in the moment of comprehension) severed or at least truncated. The "hold" of "holding together" reduces the inside under the pressure of the requirements of unity and, simultaneously, marginalizes the possibilities of relation as it tears an object out of the more indeterminate flow of inner sense.

Kant writes that "this violence must be the more significant the larger the quantum . . . comprehend[ed]." It seems that, for Kant, the experience of the sublime, in stretching the transgressive powers of comprehension to their limit and beyond, increases both the level of violence of comprehension and our sense of this violence insofar as reduction (so often accomplished effortlessly and without awareness) is felt as a rather self-conscious and failing exertion. In this sense, the sublime, as "violent to our imagination" (245; 99) is a kind of resistant counterviolence to the imagination's own violence: the presence of incommensurability and contingency in perception.

Yet, if the comprehension intrinsic to the genesis of a more determinate world of experience is always accompanied by a certain violence, that comprehension is by no means *only* transgressive. Indeed, in Kant's discussion of genius and aesthetic ideas the very same transgressive reductions through which the world is born also provide sites for creative exploration and for articulation of possibilities at the limits of the comprehended world and experience. These sites are potentially the sites of seductive pregnant absence that might animate life, language, knowledge, art, and (my extrapolation) an ethical sensibility.

Kant's discussion of genius might at first sight appear to be an ultimate exemplification of solarity (in Nietzsche's sense). On a closer reading, though, it is a story of the entwinements of giving and receiving. But first a caveat. The universalist implications of my discussion of Kant are rooted in a sense that genius and the quality of exhibiting aesthetic ideas are widespread, at least *in potentia*, among us.[61] Nietzsche expresses this idea when he writes: "Could it be that in the realm of the spirit 'Raphael without hands,' taking this phrase in the widest sense, is perhaps not the exception but the rule? Genius is perhaps not so rare after all—but the five hundred *hands* it requires to tyrannize the *kairos*, 'the right time,' seizing chance by its forelock."[62]

In other words, genuis lies dormant in most of us, waiting for the proper circumstances and animation to manifest itself. Giving and receiving among people aims to explore the possibilities for soliciting and cultivating that genius which has not yet emerged, even in places where it is least expected.

"*A* genius," Kant claims, is "a rare phenomenon" (318; 187, my emphasis). Yet Kant notes the inclusiveness of the Latin-root "genius" which accompanies "each person," and he frequently uses an indefinite and seemingly inclusive "we" when discussing the theme. Perhaps Kant thought that genius as a capacity to give and receive in this animating play is widespread though not equally spread. Most people have genius, but there are few (for all sorts of reasons) who develop fully into "geniuses." Even for those who manifest no spirit at all, however, respect engendered by the following narrative arises from a sense of the indeterminacy of genius, tied to the essential uncertainty and unpredictability of its arousal. The absence of genius can no more be described with certainty than can the rules of its presence.

Fine art, art whose "purpose is that pleasure should accompany presentations that are *ways of cognizing*" (as opposed to pleasures that are purely sensuous) (305; 172) is, as Kant writes in the heading of section 46, "the art of genius." "*Genius* is the talent (natural endowment) that gives the rule to art." Yet this giving is not, as the parenthesized "natural endowment" already indicates, that of an autonomous legislator. Rather, "since talent is an innate productive ability of the artist and as such belongs to nature, we could put it this way: *Genius* is the innate mental predisposition (*ingenium*) *through* which nature gives the rule to art" (307; 174). A fine artist can give something original and exemplary because she or he receives an endowment from nature, "*owes* a product to his genius" (my emphasis). Indebted to something received, the artist's creative activity is not and cannot become purely self-transparent (in contrast to the activity of the lawgiving author or, according to Kant's misunderstanding, the scientist). This is "presumably why the word genius is derived from [Latin] *genius*, which means the guardian and guiding spirit that each person is given as his own at birth, and to whose inspiration those original ideas are due" (308; 175).[63]

How are we to understand this receiving, this endowment from nature, and the rule that is given to art? It is certainly *possible* to interpret Kant's discussion of genius within the paradigm of sovereignty.[64] Yet, in contrast to his suggestions of a radicalized solarity—a natural being so outpouring and spontaneous that it cannot recover or grasp itself—Kant, through an understanding of the agonistic entwinement of giving with a radicalized receptivity, transfigures the conditions of possibility of giving. In an exploration of the "endowment" and the "rule given to art," the centrality and nonsovereign character of receptivity receives its first illumination. Kant claims that we all know of art, occasions, and people that lack spirit. Genius, in contrast, concerns the reception, being, and giving of spirit. Indeed, "*Spirit* [*Geist*] in an aesthetic sense is the animating principle in the mind." Yet spirit is neither self-animating nor an originality removed from relations with otherness. Rather, it is an "animating *principle*," an "ability" or "talent," a "way of offering or expressing," that at once emerges from and contributes to *a relation to the world* that is animating. As an animating *principle*, genius directs our attention and practice to this animating *relation*. Genius is at once born in relation to, and has a capacity for exhibiting, presentations that, in their vibrant ineffability, their inexhaustible proximate distance, enliven both the artist and the audience in their relations to that which is presented (313–14; 181–82).

In contrast to the autonomous will of Kant's moral theory—a will that gains neither its aims nor its power from relations to anything beyond itself—spirit, the animating principle of the mind, only "animates" in a circular relation of following and establishing the material that it presents/is presented with. Thus, Kant immediately qualifies his thoughts on spirit as animating principle:

> But what this principle uses to animate [or quicken] the soul, the material it employs for this, is what imparts to the mental powers a purposive momentum; i.e. imparts to them a play which is such that it sustains itself on its own and even strengthens the powers for such play. (313; 182)

Kant elaborates:

> Now I maintain that this principle is nothing but the ability to exhibit *aesthetic ideas*; and by an aesthetic idea I mean a presentation of the imagination which prompts much thought, but to which no determinate thought whatsoever, i.e., no [determinate] *concept*, can be adequate, so that no language can express it completely and allow us to grasp it. . . . An aesthetic idea is the counterpart (pendant) of a *rational idea*, which is, conversely, a concept to which no intuition (presentation of the imagination) can be adequate." (314; 182)

Spirit "uses" an object to animate the soul in such a way that in relation to it our mental powers are given—receive—a "purposive momentum," an activity that sustains and intensifies these powers. In relation to this other, spirit is at once able to give and receive movement and power. Yet we would profoundly misunderstand Kant here if we were to think of "material" or "employment" simply in terms of identity or presence. Pure presence would lack animating spirit: it would be the death of the soul. Rather, what spirit "exhibits" in its "presentation" is an excess that is always significantly beyond our comprehension and grasp. Spirit exhibits "aesthetic ideas" whose being consists in the way in which the inextinguishable and inexhaustible horizons within and around a presentation are made central and vital, soliciting and seductive. What is presented draws from us an infinite sense of, a striving toward, and an opening onto the not-yet-present.

Thus, "employment" here refers to the way in which spirit gives an expression that, in turn, in its very nonidentity with the self's thought, will "prompt," arouse, quicken, and animate it. Genius, spirit, is the ability to give and receive in this nonidentical entwinement through which we come to life. In this entwinement we are prompted to take up an unending effort to express "what no language can express completely," to be adequate to that to which "no determinate concept can be adequate." We are called, then, to an impossible identification. Genius is, for all its prolific *presentations*, very significantly the ability to make us aware of the inexhaustibly protean not-yet-present which is the soliciting and arousing horizon of its efforts. Genius, moreover, is the ability to be further engaged by this display. This continual emergence of possibility out of an intransigent yet seductive impossibility is the very life of genius.

This idea that the interrogative aura is both a condition for and a telos of genius becomes clearer in the light of our reflections on the transgressive reduction involved in the comprehensive birth of determinate experience. We have considered this reduction both "within" the object of experience ("running through" the manifold in light of the requirements of a determinate ordering which allows it to "hold together"; canceling temporal flow to constitute simultaneity) and in the relation between the object and its "outside" (severing possibilities of relation; resonant qualities; concealing the permeability of boundaries). Aesthetic ideas illuminate the contingent reductive nature of all determinations and draw us into an endless elaboration of concealed possibilities and concealment as such. Aesthetic ideas ceaselessly illuminate and solicit movement beyond the reductions within determinate concepts and experiences:

Now if a concept is provided with [*unterlegen*] a presentation of the imagination such that, even though this presentation belongs to the exhibition of the concept, yet it prompts, even by itself, so much thought as can never be comprehended within a determinate concept and thereby the presentation aesthetically expands the concept itself in an unlimited way, then the imagination is creative in [all of] this and sets the power of intellectual ideas (i.e., reason) in motion; it makes reason think more, when prompted by a [certain] presentation, than what can be apprehended and made distinct in the presentation (though the thought does pertain to the concept of the object presented). (314–15; 183)

Aesthetic ideas are provocatively infinite yet "pertain to the concept of the object" because all objects harbor within themselves a profusion of concealed and hitherto-transgressed possibilities. Juxtaposed to the internal unfolding of "aesthetic ideas" are presentations that provoke the elaboration of possibilities at and beyond the reductive borders of an object: "aesthetic attributes," "presentations of the imagination expressing a concept's implications and its kinship with other concepts." These presentations give "something that prompts the imagination to spread over a multitude of kindred presentations that arouse more thought than can be expressed in a concept determined by words." In "opening up . . . a view into an immense realm," they "quicken" the mind at the very edge of its violent reduction. Yet insofar as our further efforts to grasp are of necessity accompanied by further transgressions, the "immense realm of the kindred" remains essentially inexhaustible and illuminates the latter *as such* (315; 183).

If some of the examples offered by Kant concern expressions of overflowing solarity, they also tend to conceal the extent to which the themes developed here have slipped irreversibly beyond solar bounds.[65] Far more telling is the footnote which concludes the examples: "Perhaps nothing more sublime has ever been said . . . than the inscription above the temple of Isis (Mother Nature): 'I am all that is, that was, and that will be, and no mortal has lifted my veil'" (316; 185). Here Kant expresses a central agonism at the heart of the relation between "mortals" and a nonidentical world (which includes ourselves). Yet this agonism is not one of sealed defeat, but one through which an infinite articulation of selves and experiences can unfold: we are drawn to it as the site of possibility in impossibility.

This relation of giving and receiving between self and the world is entwined with analogous relations among selves' powers of imagination, understanding, and reason.[66] The aesthetic idea and the rational idea are "counterparts" in the sense that each signifies and exceeds the other, endlessly setting the other in motion. Presentations of the imagination exceed the proliferation of the determinate concepts they provoke. Presentations evoke rational ideas which exceed any intuition and again challenge the imagination to proliferate presentations in such a way that, in turn, "the power of reason" is "*set in motion*," "made to think more" than can be determinately grasped in the presentation (315; 183).

As an agonistic relation, the aesthetic idea is a series of distances that signify both a falling short and a pregnant possibility (e.g., those between the rich canvas or poem and the proliferating imagination, between this imagination and the multiplicitous grasping concepts, between reason and its idea of completeness and the ceaseless yet never complete expansion of the imagination and thought, between the richness of the latter and the abstract emptiness of reason) in the midst of which power and mental activity are enhanced. Through these distances, we are animated to a ceaselessly renewed birth—but also to endless deconstructions of our attempts to grasp completely. The "rule genius gives to art" is its particular style of evoking this interplay between presence and absence, expression and ineffability, birth and death.

In addition to emphasizing the entwinement of giving and receiving presentations, Kant's writing opens onto the vital importance of receptive relations with others for the very being of genius. It is initially through an encounter with an other that the activity of genius is animated. Each person with genius begins as an "other genius" who must receive and be swept up by the animating vortex of an other's presentation in order to give exhibitions. "The other genius, who follows the example, is *aroused by it* to a feeling of his own originality, *which allows him* to exercise in art his freedom from the constraint of rules, and to do so in such a way that art itself acquires a new rule by this" (318; 187, my emphasis. Cf. 309; 177). Thus, spirit as the animating principle must first be *animated* by an other. Insofar as each person with genius is brought to life by others, there must be "schools" and a vibrant cultural life through which ideas are nurtured to enliven "posterity" (310 and 318–19; 177–78 and 187–88). These models provide "an example that is not meant to be imitated [in the sense of copying what is present, for this is inimical to spirit] but to be followed by another genius" (318; 186–87). By engaging the inexhaustibility of an other's presentation, one is drawn into—follows—the activity of spirit and is aroused to a feeling of originality. And it is easy to see how. For to truly experience an aesthetic idea *as such* is to be animated in an activity of thought and imagination at the edge between identity and nonidentity that is so closely connected to artistic creativity. To *experience* an exhibition of another genius—as opposed to "attending an exhibition" or "copying"—is already to be aroused in a manner that significantly manifests the unique particularity of one's own talents.

The exclusively identifying mind reduces experience to an imitation (or "gives law" in the form of an autonomy that, in spite of itself, imitates the essence of imitation as transparent identity) wherein spirit "would be lost" (318; 187). In contrast to this identity—of the slave and the sovereign—Kant's account of genius decenters us and locates the animation of our mental powers in a dialogical giving and receiving within ourselves and with others. This differential arousal and transgression "makes us add to a concept much that is ineffable, but the feeling of which quickens our cognitive powers and connects language, which otherwise would be mere letters, with spirit" (316; 185). Language *as language*, as opposed to mere letters, has its very being in the opening movement between identities and the nonidentical, in the reciprocal provocations within the self and between self, others, and otherness.

Kant claims that genius exhibits presentations of fine art in large part to arouse the *pleasure* associated with the quickening of the cognitive powers that accompanies the reception of such presentations. However, this claim clearly does not exhaust his account of aesthetic ideas, for aesthetic ideas ceaselessly outstrip and defeat the understanding's efforts to comprehend the presentation completely. This striving activates reason to "think more," but the "more" is attributable in part to an ineliminable sense of the "less" which characterizes understanding and reason in comparison to the *open* proliferating infinity prompted by the aesthetic idea. Hence it appears that the emotional experience of aesthetic ideas is closer to that of the sublime with its combination of pleasure and pain.[67] Yet here there would be an endless oscillation between the two emotions, rather than a culmination in pleasure stemming from reason's triumph.

Even with this transfiguration of Kant's description of the emotions connected with aesthetic ideas, it is crucial to recognize that this animation of the self is by no means restricted to the aesthetic dimension concerning pure form, formlessness, pleasure, and pain. Rather, the aesthetic is intertwined with the epistemological. Thus Kant writes that when the understanding is animated by an aesthetic idea, it "employs material not so much objectively for cognition, as subjectively, namely, to quicken the cognitive powers, though indirectly this does serve cognition too" (317; 185). Similarly, Kant discusses the poet who "announces merely an entertaining *play* with ideas, and yet the understanding gets as much out of this as if he had intended merely to engage in its own task" (321; 190).

Even these statements, however, do not go as far as Kant's analysis would suggest. Cognition of the object is not simply an *incidental* aspect of aesthetic experience and presentation. Rather, it is integral to it insofar as these presentations significantly concern the elaboration of the content of the matter at hand, evoking and exploring myriad aspects and possibilities within a concept or object, or in relation to "kindred presentations" that exceed the reductive grasp of any single comprehension. The subjective play of the cognitive powers comes to life through substantive engagements with hitherto-suppressed possibilities for expressing (and simultaneously transgressing) a determinately given object. Integral to aesthetic animation is that thought must "think more" about the object, and come to a more explicit cognition of the object—and our relation to it—as essentially inexhaustible. At the same time, this subjective "quickening" and the feelings of pleasure and pain involved, react back on cognition, pushing it beyond its present limits, opening new efforts and possibilities for understanding.

Thus while Gadamer's criticism of Kant's subjectivization of aesthetics is generally correct, we can see in this significant backwash in the *Critique of Judgment* an entwinement of aesthetics and cognition that markedly and suggestively contrasts with Kant's dominant themes. Now, if we understand the genesis of cognition in relation to an aesthetic sense of the inexhaustibility of a world that solicits and thwarts our cognitive strivings—and if we understand the elaboration of these strivings as entwined with and animated by a play between pleasures of expansion and pains of inadequacy—if *all* cognition, as movement beyond the violent reductions accompanying each determinate experience, has these characteristics (not just the cognition provoked by aesthetic ideas where these characteristics are most evident), then we find ourselves on a terrain where self and experience emerge not through subjective legislation, but through an open and unending articulation.

This terrain involves aesthetically infused possibilities whose shapes are forged along crooked paths of expansion and retreat in relations of giving and receiving with a nonidentical world. The unities of self and experience are continually made, unmade, and remade in this ceaseless gathering and releasing of that which is never completely amenable to any determination. From this perspective, intelligence itself would be significantly rooted in an indeterminate play among selves and between selves and the world.

Of course, Kant does not pursue these thoughts in the Third Critique, as is clear when he claims that Sir Isaac Newton's discoveries, like all *cognition*, lie entirely "in the natural path of an investigation and meditation by rules and [do] not differ in kind from what a diligent person can acquire by means of imitation" (308; 176). Though it took a "great mind" to make Newton's discoveries, Kant views them as the outcome of a methodically governed progression that differs from the everyday imitator "only in degree" (309; 177).

This understanding of cognition, however, makes sense only if Kant successfully deduced the continuous order of experience from a subject which gives the laws to nature. Yet Kant failed here. If instead there is fundamental contingency, reduction, and transgression entwined with the birth of all objects of experience, then the idea—even the *"regulative* idea"—of a seamless and continuous rule-governed order of things makes little sense; nor can we make much sense of the corresponding idea of inquiry.[68] Rather, while understanding can be extended through the development of logics contained and implied in what is already determinately given, we are simultaneously called to explore and discern that which would be discontinuous with what is given, that which is powerfully concealed by extant determinations and by all that seems possible within their framework. This latter, wilder task involves a more agonistic, playful, and painful encounter (a less determinate encounter) with the at once soliciting, recalcitrant, and inexhaustible horizon at the limit of our determinations. Without such an engagement, our thinking and experience begin to lose their vitality and intelligence, and language becomes "mere letters."

If we bear all this in mind and further underscore the importance of relations among selves in animating and sustaining this wilder encounter, then some of Kant's most intriguing remarks on the public sphere in "What Is Enlightenment?" come into focus in a highly suggestive manner. Kant's opening is well-known: "*Enlightenment is man's emer-gence [Ausgangs: exit, way out, release, escape] from his self-imposed imma-turity [Unmündigkeit]. Immaturity* is the inability to use one's under-standing without guidance from another. . . . [It is] self-imposed when its cause lies . . . in lack of resolve and courage to use it without guid-ance from another."[69] Interestingly, Kant does *not* locate the likelihood of enlightenment in the efforts of atomistic individuals to think freely, for these efforts are likely to be too feeble and insecure. Rather, "that *the public* should enlighten itself is more likely; indeed, if it is only allowed freedom, enlightenment is almost inevitable" (36; 41–42, my emphasis).

Yet what is it about the free public use of our cognitive powers "be-fore the entire literate public" that leads Kant to be so optimistic, espe-cially in light of his strong critical awareness of widespread laziness, cowardice, socially engendered stupidity, and fear? Does Kant have in mind the expansive universalist attitude that one must adopt as one argues before everyone else? Does this attitude draw us out of our-selves such that, in public, thus drawn, we elaborate the very essence of "understanding"? This Habermasian reading has substantial merit, and I shall return to it shortly. But Kant's essay strikes—perhaps even *ac-cents*—another chord as well.

Missing in the strictly consensualist reading is the emphasis Kant places on the self's *distinctness* in its enlightened use of the cognitive faculties: a use "without guidance"; "in accord with his own lights," with "each person's calling to think for himself." The importance of the specificity of this effort, as well as the proximity between this emphasis and Kant's discussion of cognition in relation to genius and aesthetic ideas, is evoked in the following: "Rules and formulas, those mechani-cal aids to the rational use, or rather misuse, of his natural gifts, are the shackles of a permanent immaturity" (36; 33). Here Kant views the idea that thinking is wholly continuous with given rules as erroneous and constitutive of the "shackles of a permanent immaturity," insofar as the essence of thinking lies in the repeated encounter with an inexhaustible manifold world which is always significantly beyond one's grasp—be-yond the rules. The effort to use one's cognitive powers involves ex-ploring possibilities of perceiving and understanding which lie at and beyond the limits of extant perceptions and formulas. This effort de-pends on a courage to engage the specific and protean nexus which lies between (and is in turn constitutive of) the self and the surrounding world. This receptive engagement requires a *moment* of commitment to that ineffable "multiplicity of partial representations" which exceeds concern for public acceptability.[70]

Yet Kant does not confuse the utter *distinctness* of each self's perception and voice with an atomistic rendering of their conditions of possibility. Rather, concerning the latter, he echoes and emphasizes the above-discussed themes having to do with the mutually animating quality of selves' encounters with other selves similarly engaged. Kant expresses hope in such an enlivening public when he writes of "a few who, having thrown off the yoke of immaturity, will spread the spirit of a rational appreciation for both their own worth and for each person's ability to think for himself" (36; 42). This spread of "spirit," via the engagement with another's soliciting efforts to think beyond the rule, is further articulated in the closing sentence of the essay when Kant writes: "Once nature has removed the hard shell from this kernel for which she has most fondly cared, namely, the inclination to and vocation for free thinking, the kernel gradually reacts on a people's mentality" (41; 46). A significant part of this reaction is that of a solicitation in the face of the inexhaustible, an empowerment of reciprocal provocation, an intoxicating "quickening" of one's cognitive powers and being, a compelling alternation between pleasure and pain in the endless advances and retreats with nonidentity.

If we were to stop here, however, we would fall far short not only of what Kant offers but of the phenomena at hand. Kant's understanding of the free public use of our cognitive powers seeks to animate not just the interplay of distinct beings striving to speak and hear beyond the rule, but also a will to present and judge offerings "before an entire literate world." Here the aim of "arguing" is clearly to establish a unanimity to which each, using his or her judgment, can assent. This theme is articulated in Kant's numerous essays on the public sphere and is perhaps most succinctly formulated when he writes, in the *Critique of Pure Reason*, of a "criticism [to which] everything must submit" which opens the possibility of "the sincere respect which reason accords only to that which has been able to sustain the test of free and open examination" (Axi). One should fashion and judge claims in light of the idea of a depth and scope that is worthy of universal assent. In related reflections on taste, Kant writes in the Third Critique of a *sensus communis*, which means the idea of a sense *shared* [by all of us], i.e., a power to judge that in reflecting takes account (*a priori*), in our thought, of everyone else's way of presenting [something], in order *as it were* to compare our own judgement with human reason in general and thus escape the illusion that arises from the ease of mistaking subjective and private conditions for objective ones, an illusion that would have a prejudicial influence on judgement. . . . [We] compare our judgement not so much with the actual as rather with the merely possible judgements of others, and [thus] put ourselves in the position of everyone else, merely by abstracting from the limitations that may happen to attach to our own judging. (293–94; 160)

Writing on the public sphere, in "What Is Orientation Thinking?" Kant argues that the idea of reasoning before a communicating community is essential to prevent genius from becoming so "delighted with its daring flights" that it sinks into unchecked "superstition" and "zealotry."[71]

This concern for the "universal standpoint" shared by taste and understanding follows a path of consensus-driven engagement with actual and possible others, an engagement that establishes intersubjective continuity and intrasubjective consistency. The hope is to check the dangers of subjectivism and oblivious fanaticism. Consensualist readings of Kant follow and accent his dominant voice, and thus they often smother the wilder and more differential voice in the consensualist breath of the *sensus communis*. Yet this too misses a vital possibility.

Kant often suggests that a combination of (and an agonism between) differential genius and consensual judgments of taste is vital for the production of fine art. Of course, where the two are incommensurable, he leaves no doubt which must reign: "If there is a conflict . . . and something has to be sacrificed, then it should rather be on the side of genius." This sacrifice is metaphorically elaborated in a variety of ways:

Taste, like the power of judgment in general, consists in disciplining (or training) genius. It severely clips its wings, and makes it civilized, or polished; but at the same time it gives it guidance as to how far and over what it may spread while still remaining purposive. It introduces clarity and order into a wealth of thought and hence makes the ideas durable, fit for approval that is both lasting and universal, and hence fit for being followed by others and fit for an ever advancing culture. (319–20; 188)

This a priori privileging of judgments of taste over genius is based on a transcendental theory of a *sensus communis* grounded in the subject's universal cognitive faculties. But if this theory succumbs to a critique of the sovereignty narrative, then taste itself and its relationship with genius opens to fresh questioning.

Without transcendental universalism, the activity of consensualist judgment still has an important function in cognition and human interaction. We can consider this significance in two ways. The first concerns the exercise of our cognitive powers in relation to an inexhaustibly manifold and nonidentical world. If we understand this exercise as an effort to go beyond the transgressive reductions that accompany the birth for us of all determinate objects and concepts, then the demand to consider our understanding and judgment from a perspective that takes account of "everyone else's way of presenting" appears as a call to inquire into the "limitations" and "illusions" that cling to our own and other particular perspectives in the critical lights of actual-specific and imagined-general others. In this sense, the regulative idea of giving or judging a presentation in a manner worthy of consensual embrace generates an interrogative illumination that might open us to and beyond our extant limitations in a manner consonant with the spirit of genius. Involved here is a striving that presses questions on us concerning possible relations, overlaps, and supplements among diverse visions, not to mention negations.

The second sense in which a significantly consensual judgment remains vital in the absence of transcendental universalism is related to the dialogical conditions for the activity of genius itself. It is not that the dialogical activity of genius is essentially consensual. Rather, we have seen that genius emerges and thrives on agonistic solicitations and resistances among the cognitive powers within the self, between the self and the world of aesthetic ideas, and among selves. Yet insofar as selves reflect back on the dialogical context of their enlivening, they find themselves profoundly indebted to this social context of inexhaustible others. Might not this profound debt be intertwined with a profound generosity that strives to receive these others *as others* and to give a presentation that might provoke an enlivening reception on the part of the others? Clearly, giving in this latter sense involves the effort to present a difference that provokes. Yet the concern to provoke is always attached to a concern to *communicate*, and always harbors an aspect of consensuality—not in the thick sense of striving toward an exhaustive agreement, but in the thinner sense of striving to establish enough shared understandings to enable the others to be swept into the materiality and sense of what is presented, so that others open onto the profundity of another perspective with all its inexhaustible representations, directions, and questions. The concern to offer enlivening alterity is always entangled in the game of establishing overlaps through which the giving can become a "gift" and not, as Kant says, "nothing but nonsense" as far as the others are concerned (319; 188).

Hence there is, for reasons concerning both the development of cognition and the generous relation to that human plurality which is the condition of possibility of thriving, something important and legitimate about the consensual concerns of taste. These concerns are partly in tension with the wilder concerns and engagements with multiplicity and inexhaustibility peculiar to genius, and thus Kant is not wrong to describe taste's relation to genius as one of "discipling," "clipping wings," "civilizing," and "giving guidance," as "introducing clarity and order" so that genius can be "fit for being followed by others." The importance of consensuality, however, does not establish the universal hegemony of taste over genius where they seriously conflict; and this is so not simply because the limits and possibilities of communicability are highly ambiguous. More important, the animating enlightenment of selves and communities is most probable precisely in the *agonism* whereby these two concerns critically open each other and us to a profusion of presentations and questionings that solicit a generous and receptive freedom. Participation in the public sphere ought to consist in an endless effort to negotiate this tension, not to collapse it.

If we sever the *sensus communis* from the a priori hegemony of taste and from the singular concern for (and understanding of) the "shared" in terms of consensus, we might reappropriate this term—this "common sense" which consists in a sense for the common—as a cultivated sense of the *tensions* between identity and difference on which the generous and receptive vitality of selves and communities depends. It is a sense that can be cultivated only through active participation in the life of sharing spirit with others; a sense that is at least slightly transfigured with each new voice and ear.

Emerging in this discussion is a transfiguration of a certain respect for persons, albeit by way of a narrative very different from Kant's. Respect no longer stems from the capacity of each person unconditionally to reject relations with the world as the determining ground of the will. Instead, it emerges out of a sense that the freedom of new beginnings, possibilities, and insights is animated in the mutual provocations of giving and receiving. This freedom is the product not of an essentially solitary sovereignty but of relations of arousing (and being aroused) through the presentations we offer to (and receive from) each other in a dialogical context. The respect drawn in this context is a respect for each person's distinct yet indeterminately open capacity to partake in these relations. Would not a respect thus understood be more likely to be intimately concerned with the specificities of the world and the others with whom we are entwined?

I began by recounting some of the difficulties modernity faces in imagining the possibility of reception in a world significantly withdrawn from the light of a generous God. Kant's practical imperative not to treat reception as foundational exemplifies the fundamental suspicion with which modern selves confront their world. Fundamental receptivity, Kant thought, invaded the autonomy of both self and other; it reduced all relations with otherness to self-love, and reduced self-love to contingency. In an eddy, Kant imagines another kind of reception—another desire, another pleasure and pain—which reaches out and opens up in order to sense and generously engage something that transcends the reductions of self-love. Here, giving hinges on receiving others and otherness with a respect that honors an inexhaustible distance. One could say that modernity's distance from the world increases in direct relation to the distance of a benevolent God's retreat. One strange response to this distance is to seek to conquer it by convincing ourselves that we are sovereign and autonomous subject-gods. But this is a doubtful project. Another strategy, the "eddy strategy," is to inhabit this distance as a space of highest possibility. Yet to pursue the latter path in the midst of a world that travels largely on the former is to confront a project that is not nearly so "happy" as Kant's description of the play of genius. To explore further the possibility of an ethics of receptive generosity, to embrace the latter ethics as at once hopeful and tragic, I now turn to the work of Theodor Adorno. I read Adorno partly animated by a certain cheerfulness in Kant, even as I seek to infuse Kant's more hopeful directions with a greater sense of the tragic.

CHAPTER TWO

Adorno: Toward a Post-Secular *Caritas*

My discussion of Kant in Chapter 1 illuminated the epistemological, ethical, and political incoherence and danger within modernity's most powerful articulation of sovereign subjectivity. Part of why Kant's corpus is compelling is the intellectual rigor with which he negotiates a razor-edged path: one eye on the abyss that appears wherever the sovereign subject is lacking, the other eye on the abyss that appears when subjectivity is affirmed in the absence of all receptivity. If this strange journey appears largely untenable, nevertheless Kant's errors illuminate far more than most theorists' truths.

Wishing to depart from the Kantian path toward more promising terrain, I interpreted parts of Kant's Third Critique that suggest an alternative formulation of giving and receiving as the birth of sense, ethics, and the public sphere. Now, in order to deepen and broaden these ideas, I turn to the thought of Theodor Adorno, for whom the reconceptualization of giving and receiving is central. Since my interpretation of Adorno differs from the dominant readings of his work, I shall introduce Adorno by reciting two schools of critical interpretation. I focus primarily on Habermas's interpretation, which has received substantial support from contemporary critical theorists, and then sketch a postmodern critique of Adorno. My summary of these readings will provide a context within which to distinguish themes that facilitate an engagement with Adorno. In section two, I develop a reading of the *Dialectic of Enlightenment* that draws out the "positive notion of enlightenment" as receptive generosity, a notion which operates throughout that text and which is developed in the rest of this chapter. Section three addresses Adorno's "morality of thinking" by exploring his frequent comments on giving and receiving, and section four elaborates the ways in which Adorno sought to articulate an ethics by positing a constellation of solicitations in contrast to rule-governed morality. In the fifth section, I follow the dialogical contours of Adorno's thinking and discuss his powerful critique of "what is called communication." (That critique, I argue in Chapter 3, has substantial import for the Habermasian paradigm.) I shall conclude this chapter with my reflections on Adorno's capacity to respond to questions raised by Charles Taylor concerning the possibility of nontheological moral sources.

Adorno and His Interpreters

One might formulate Habermas's critique of Adorno thus: Adorno seizes, and is seized by, the philosophy of sovereign subjectivity and pushes it self-critically until it shatters into a million incoherent pieces. Then, draped in black, he refuses to leave the space of the subject whose death he pronounces. If reconciliation and freedom cannot be accomplished by a subjectivity whose epistemological and ethical giving is revealed as other-denying rage, nevertheless this subject cannot be given up. Yet instead of being the origin of truth, being, and the moral law, the shattered subject's only remaining gift—both legitimate and imperative—is to give death to its *deadly giving itself* in an endless negative dialectic. If subjectivity ragefully subjugates the surrounding world and its own embodied being, then the only remaining task for one still animated by the ungrounded injunction to recognize otherness as an end in itself is that of continually robbing the robber: blowing the whistle louder and louder into the ears of a subjectivity that does not want to hear it. Adorno writes not because this course of action will help us, for we are incapable of receiving the other such that we might truly give. Rather, he writes early on to a merely "imaginary witness" (*DE*, 256) and, a little later, "only for the dead God" (*MM*, 209).

Of course, Habermas does not claim that Adorno was philosophically worthless. Indeed, Adorno remains the most brilliant person he has ever met.[1] For Habermas, Adorno's work is an enlightening error that tacitly points toward the communicative rationality to which

Habermas has devoted much of his life. What seems to sparkle in Adorno is, paradoxically, the integrity of his account of the night.

On a scale of luminosity, Habermas seems to locate Adorno and Horkheimer somewhere between "dark" (Machiavelli, Hobbes) and "black" (Sade, Nietzsche). And the *Dialectic of Enlightenment*, says Habermas, is "their blackest book." Abandoning "hope in the liberating force of enlightenment," it is nevertheless a book "unwilling to relinquish the paradoxical labor of conceptualization" (*PDM*, 106).

While the Marxist critique of ideology differentiates between knowledge and power, and undermines false claims to the former by showing them to be sustained only by the latter, Adorno and Horkheimer (according to Habermas) undermine even the privileged position of ideology critique by turning the suspicion of the power-drenched bankruptcy of truth back on itself in a totalizing self-destructive manner.[2] However, in contrast to Nietzsche and to twentieth-century Nietzscheans who develop totalizing critiques of reason and then attempt to deny the "performative contradiction" involved in making validity claims while denying the legitimacy of such claims, Adorno's totalizing critique stands resolutely in the face of this contradiction. Thus, "*Negative Dialectics* reads like a continuing explanation of why we have to circle about within this *performative contradiction* and indeed even remain there" (*PDM*, 119). Without hope in the enlightenment he will not surrender, Adorno permits reason to "shrivel" to a mimetic impulse that must—but cannot—be recovered. "In the mimetic powers the promise of reconciliation is sublated. For Adorno that then leads to *Negative Dialectics*—in other words to Nowhere" (*AS*, 90). Without rational normative foundations, critique defeats itself.

"How," Habermas wonders, "can these two men of the Enlightenment . . . be so unappreciative of the rational content of cultural modernity that all they perceive everywhere is a binding of reason and domination?" (*PDM*, 121). In Habermas's view, it is because, even as unyielding critics of subjectivity as "self-preservation gone wild," they (most resolutely, Adorno) remain within this paradigm and are thus limited to "two attitudes of mind": namely, "representation and action." Our representations are fundamentally tied to the possibility of instrumental efficacy, which in turn requires this knowledge. Even critical theory remains within this blind and violent structure and, hence, cannot articulate any guiding notion of reconciliation, for that would require an impossible access to something beyond instrumental reason.

Yet the very reconciliation that is *inconceivable* is simultaneously *exagger-ated* and *demanded* by the Nietzschean aesthetic subjectivism by which Adorno and Horkheimer "let themselves be inspired" (*PDM*, 121). This aestheticism is the most extreme manifestation of presentist time-consciousness, of modernity's radical break with all consciousness rooted in tradition, to the point where the most transitory self-given moment becomes supremely important for our interpretation of (and relation to) time and being. Such an aestheticism overwhelms norma-tive questions with a "longing for an unspoiled inward presence" (*PDM*, 123), through which *all* practices necessarily appear subjugative. Permanently revoked and radically invoked, Adorno's reconciliatory impulse remains far beyond the discursive realm in a (philosophically useless) presupposed "original relation of spirit and nature [that] is se-cretly conceived in such a way that . . . truth is connected with . . . universal reconciliation—where reconciliation includes the interaction of human beings with nature, with animals, plants, and minerals" (*TCA* 1, 381). Adorno, in short, conjures up "the utopia of a long since lost, uncoerced and intuitive knowledge belonging to a primal past" (*PDM*, 186).

Yet Adorno appears as a *relatively* good philosopher when compared to the other post-Nietzscheans discussed in *The Philosophical Discourse of Modernity*. For the integrity of Adorno's thought leads him to remain within (and develop the contradictions of) epistemological, normative, and aesthetic monological subjectivism. By so doing, Adorno makes problems visible that lesser philosophers conceal, and he thereby "fur-nishes us with reasons for a *change in paradigm* within social theory" (*TCA* 1, 366). Indeed, in response to Adorno's elusive passage about how "the reconciled state . . . would find its happiness in the fact that the alien remained distinct and remote within the preserved proximity, beyond being either heterogeneous or one's own" (*ND*, 191), Haber-mas writes: "Whoever meditates on this assertion will become aware that the condition described, although never real, is still most intimate and familiar to us. It has the structure of a life together in communica-tion that is free from coercion. We necessarily anticipate such a reality . . . each time we want to speak what is true. The idea of truth, already implicit in the first sentence spoken, can be shaped only on the model of the idealized agreements aimed for in communication free from domination" (*PPP*, 108–9). In short, Adorno clarifies the insurmount-able difficulties of subjectivism and unwittingly gestures in directions that, as "whoever meditates will become aware," lead toward Habermas.

Adorno's reception by what may loosely be called postmodernism has been mixed.[3] Many postmodernists see much that is admirable and worthy of further development in Adorno, often seeing him as a precursor of their own positions. Yet at the same time, many of them criticize Adorno for being unable to break fully with modernism on questions of truth and subjectivity. Most of this group resist those aspects in Adorno which Habermas respects, and are attracted to those aspects Habermas most resists.

Harmut and Gernot Böhme, in *Das Ander der Vernunft*, state their own position succinctly, writing that in Horkheimer and Adorno's theory, "reason is criticized as instrumental, repressive, narrow. . . . Their critique takes place in the name of a superior reason, namely, the comprehensive reason, to which the intention of totality is conceded, though it was always disputed when it came to real reason. There is no comprehensive reason. . . . [R]eason does not exist apart from its other. . . . [I]t becomes necessary in virtue of this other" (quoted in *PDM*, 304–5). Michael Shapiro, an insightful reader of Adorno, discerns a similar problem, though he expresses it more tentatively. Shapiro suspects that Adorno falls short of Foucault's insight that "all modes of intelligibility are appropriations," for "critical theorists in general, and Adorno to some extent (although his position is ambiguous here), base their readings of the reification of the self on a model of an authentic mode of intelligibility."[4] Reacting differently to the ambiguity of Adorno's thought, Stanley Fish argues that because Adorno resolutely commits himself to the impossibility of escaping "the objective context of delusion," he must forswear any positive yield from negative dialectics, and must pursue instead a restless self-critical motion. This motion, however, is driven by a "militantly secular faith"—a baseless faith with a substanceless hope—not that critical self-consciousness can be achieved but that "some (unimaginable) force will burst in upon it." Unable to quit appealing to a vague idea of some transparent emancipatory condition, Adorno retreats into the opaque brush of "millenarian prophecy" and appeals to "a generalized human potential" that is essentially unthinkable. Yet Adorno's ambiguity is not (as Shapiro implies) attributable to philosophical uncertainties concerning "authenticity," but is rather the necessary strategic form taken by a "militantly secular faith" that is critical of all privileged positivity, yet privileges itself.[5]

These postmodern critiques share the idea that Adorno is unable to free himself from a notion of an "authentic intelligibility" that animates his textual practice. If Habermas's problem with Adorno concerns his inability to provide a notion of rational reconciliation, for postmodernism the problem concerns Adorno's inability to quit evoking such reconciliation. Many of Adorno's modernist and postmodernist critics discern a radical presentism in his work. Whether this presence is conceived of as the revelatory aesthetic flash or as an authentic intelligibility, it is frequently understood to obscure what is most important: rational communication, or the otherness that renders all discourses equally arbitrary.

What all these interpretations miss (though Shapiro is closer to the mark) is Adorno's insistence that negative dialectics is most fundamentally a performance, a happening, a textual practice. It is not a practice whose meaning would lie in some transparent presence utterly incommensurable with itself, but rather one whose meanings lie largely in the ways its very movements exemplify ethical engagement. Rather than viewing Adorno's reflections on "reconciliation" as the substantive aim of his thinking, I read them as ideas that must be understood as parts of a constellation of concepts that Adorno composes with the hope of engendering and exemplifying a particular type of thinking activity. In other words, I hold fast to Adorno's repeated insistence that "the crux is what happens in a philosophy, not a thesis or position—the texture" (*ND*, 33). Similarly, "the test of the turn to nonidentity is its performance; if it remained declarative, *it would be revoking itself*" (155). Given Adorno's claim that all declarations about the world are nonidentical with that which they seek to identify, thinking must be the endlessly renewed activity of moving beyond one's current conceptualizations. Contra Fish, Adorno's emphasis on performance stems not from a compulsory taboo against declaration—Adorno is hyper-declarative, he revels in declaration—but from his concern for the "morality of thinking" (*MM*, 73), which includes yet exceeds the declarative to manifest itself in the very *event* of thinking.

Reading Adorno this way does not mean that we should understand the "happening" or "performance" of negative dialectics as independent of, or more fundamental than, the subjects and objects it mediates. To do so would hypostatize negative dialectics in a manner analogous to (Adorno's reading of) Heidegger's "Being," which Adorno criticizes as a relation or mediation too separated from the ontic entities that mediate this mediation. Rather, negative dialectics must be an event of thinking that has a "compulsory substantiveness": the activity of conceptualization must continually exceed its concepts to engage the nonidentical subjects and objects from which it comes and toward which it moves. "The concept of nonconceptuality cannot stay with itself" (*ND*, 137). Thus, while the "crux" of negative dialectics lies in its activity, it must not fetishize and reinclose itself in a self-congratulation of the performative but, rather, interpret and enact itself as an *opening up* such that the newly identified activity "cannot stay with itself."

If these injunctions are to avoid meaninglessness, though, Adorno must offer a conceptualization—however tentative, paradoxical, and swept up in the movement of thinking—of a particular type of ethical activity and of the ways in which it is preferable to other alternatives.[6] But how to describe and animate this activity, this particular morality of thinking, that seeks to exceed the hypostatization of concepts and of activity itself?

Contrary to most interpreters, I think Adorno does offer a broad ethical perspective.[7] He understands negative dialectics' repeated and distinct outstripping—of concepts by concepts, of concepts by activity, of activity by concepts—as an activity of giving and receiving through which concealed suffering is revealed and higher possibilities are evoked and explored. Inseparable from Adorno's concern with the "morality of thinking" is his claim (animating *Minima Moralia*, which opens with a quotation of Ferdinand Kürnberger: "Life does not live") that "every undistorted relationship, perhaps even the conciliation that is part of organic life itself, is a gift" (*MM*, 43). Negative dialectics is endlessly soliciting receptive generosity. Yet every response to this call engenders identities that perpetrate, to varying degrees, a tragic erring and theft. This erring does not mean that the practice of generosity is impossible, nor that the theft inexorably outweighs or equals the gift. However, it does mean that giving falls short of the highest solicitation to give and that it proliferates dangers with each move. Hence, receptive generosity is essentially the endlessly renewed effort to give beyond the given gift.

Of course, this ethical pronouncement might wither into a mere declaration which ceases to move and be moved beyond itself toward otherness. Do not ethical *declarations as such* risk becoming reifications, which absorb us in processes of doctrine maintenance and adherence-to-the-letter that displace and eclipse the very other-respecting activities they proclaim—securing "generous identities" to the preclusion of events of receptive generosity? To negotiate both the need for and the dangers of declaration, Adorno develops his theoretical project in an essentially "constellational" mode. That is to say, he employs declarations in a constellation of concepts whose relations of tension and overlap disrupt the declarative and solicit an active morality of thinking. If one forgets the centrality of this constellational textual practice, Adorno's project becomes virtually incomprehensible.

Though I later discuss in detail Adorno's concept and practice of composing constellations, a few prefatory comments are called for. Many have noted that Adorno employs constellations of concepts as a means of illuminating those aspects of the specificity and excess of the world which elude individual concepts.[8] Adorno juxtaposes concepts not simply to create a larger sum of insights through addition but, moreover, so that their *relationships* of overlapping and tensionality can evoke that which is essentially incommensurable to the singularity of a concept or simple sum of concepts. These relations have an essentially dynamic quality, such that they *solicit thinking* as much as they illuminate particular objects of thought.[9] Thus they are vital to a mode of thinking that is at once declarative and more than declarative. This latter task opens—beyond the theoretical-interpretive dimension—an essentially practical-ethical aspect to the constellations Adorno constructs. Since ethical relations with others and with otherness cannot be exhausted by declarations, part of Adorno's project is to illuminate and solicit a diverse set of often agonistic concerns that are to animate our receptive generosity—our "morality of thinking."

Some read Adorno's project, beginning with the *Dialectic of Enlightenment*, as a hopeless performative contradiction that culminates in *Aesthetic Theory*, unfinished at his death. On this reading, negative dialectics leads to Habermas's (unhelpful) insight that "if a spark of reason is left, then it is to be found in esoteric art" (*AS*, 98). *Negative Dialectics*, "which develops the paradoxical concept of the nonidentical, points to [*Aesthetic Theory*], which deciphers the mimetic content hidden in avant-garde works of art" (*PDM*, 129). Adorno's final work thus constitutes a total retreat from theoretical modernity.[10]

Not only does this position fly in the face of the chronological fact that *Aesthetic Theory* was begun *before Negative Dialectics*, interrupted by the latter, and then resumed, but it also makes thoroughly unintelligible how and why Adorno planned to follow *Aesthetic Theory* with a work on moral philosophy that, along with the other two masterpieces, was to "represent the quintessence of my thought" (*AT*, 493–94). My own claim is that Adorno's thinking was, from early on, quintessentially ethical—whatever else it may have been at the same time. Adorno's critique of epistemology as "first philosophy"[11] opened the possibility of understanding philosophy essentially as an ethical movement beyond ourselves toward engagements with nonidentity. Though Adorno's position is substantially different from that of Emmanuel Levinas, there is a proximity in the priority each gives to ethics. Hence, I do not see what follows so much as an effort to sketch the moral philosophy Adorno never lived to write, but rather as an attempt to bring into clearer focus the morality of thinking he had always been practicing and writing. In contrast to Habermas's claim that Adorno and Horkheimer abandoned "hope in the liberating force of enlightenment," I wish to consider carefully the ethical meaning of their claim in the introduction to the *Dialectic of Enlightenment* that "we are wholly convinced—and therein lies our *petitio principii*—that social freedom is inseparable from enlightened thought." I seek to explore their idea of a morality of thinking which ceaselessly interrogates the "recidivist element" that clings to reason, in order to open possibilities of something more and better than the fateful *equivalence* of knowledge and domination (*DE*, xiii). Far from hopeless, the "critique of enlightenment," from the beginning, "is intended to prepare the way for a positive notion of enlightenment which will release it from entanglement in blind domination" (xvi). This hope, this enlightenment, precisely that of a philosophy of receptive generosity, is already at work in the critique.

I proceed with a reading of the *Dialectic of Enlightenment* that draws out the repeated glimpses Adorno and Horkheimer offer of a "positive notion of enlightenment" which succumbs again and again in history. These glimpses of possibility disrupt the Habermasian claim that this relatively early work is unable to point significantly beyond negativist despair.[12] More important, these accounts harbor a cluster of insights—concerning dialogue, difference, freedom, and generosity—that constitute the seeds of the "morality of thinking" which Adorno develops in later works. Their marginal appearance in the *Dialectic of Enlightenment* should not be read as evidence that they must *inexorably* and *totally* fall victim to and even precipitate enlightenment's self-defeating dynamic, but rather that thus far they have emerged in overwhelmingly pressurized contexts which have extinguished their more hopeful possibilities. The *Dialectic of Enlightenment* provides a series of narratives that illustrate the mutual destitution brought on by efforts to gain mastery through subjugation and the denial of otherness. In this way, the text *indirectly* urges us toward a more intertwined dialogical freedom and generosity. At the same time, it provides gestures that solicit us more *directly* toward these alternatives. We should not confuse the brevity of these gestures with the extent of their significance. In a different context, concerning Benjamin, Adorno wrote: "As in good musical variations, this theme rarely states itself openly."[13] If we do not understand this, we will understand little of Adorno.

Receptive Generosity Draped in Black

In the introduction to the *Dialectic of Enlightenment*, we encounter the central and jarring claims that "myth is already enlightenment; and enlightenment reverts to mythology" (xvi). Meant to challenge enlightenment's sense of the radical discontinuity between itself and that which preceded it, this assertion can be taken many ways.[14] For different reasons, strong proponents of the enlightenment as well as its opponents have an interest in interpreting this statement as a proclamation of a fateful *equivalence* between myth and enlightenment. This interpretation is helpful in opposing enlightenment because it suggests that the understanding of enlightenment as reason's discontinuity with, and opposition to, blindness and power is itself utterly mythical. For strong advocates of the enlightenment tradition, the appeal of this interpretation of Adorno and Horkheimer's central claims is twofold. First, it shows that their critical theory offers nothing but a self-undermining and easily dismissible critique of enlightenment. By claiming to enlighten the enlightenment about itself while simultaneously rejecting all enlightening as mythological, they succumb to the proverbial "performative contradiction." Second, the equivalence between myth and enlightenment suggests the futility of any hope for an alternative enlightenment based on their radical critique.

I illuminate these interests and appeals in order to raise a suspicion regarding the deep ruts and the overused paths followed by many interpreters of the *Dialectic of Enlightenment*. With Benjamin, Adorno notes that "the eyes of countless beholders have left deep tracks on some old pictures" (*AT*, 276). Regarding this text, I think, there has been substantial damage. Bearing in mind that the sentence immediately preceding the one on myth and enlightenment proclaims their intention to "prepare the way for a positive notion of enlightenment which releases it from entanglement in blind domination" (*DE*, xvi), I pursue a reading that shows mythical closure to be intrinsic to, but by no means totally definitive of, enlightenment. By reflecting on enlightenment's dangers, identifying the conditions most conducive to their proliferation, and searching for ways to subvert them, Adorno and Horkheimer seek to gather together insights toward an enlightenment capable of more desirable directions.

The most primordial form of the proximate relations between myth and enlightenment is articulated in Adorno and Horkheimer's discussion of "*mana* in the earliest known stages of humanity."[15] *Mana*, on their reading, is the "gasp of surprise" in the face of that which is uncommon and unexpected. It indicates "that which transcends the confines of experience; whatever in things is more than their previously known reality" (*DE*, 15).

For Adorno and Horkheimer, pre-animistic *mana* signifies both an opening for enlightening relations (defined in part as those which seek to engage the world we have identified with an awareness of the other's or otherness's transcendence of our identifications) and a mythic closure. The possibility of enlightenment dawns in the acknowledgment of an otherness, beyond the confines of our knowledge and identifications, with which our thought and being are deeply entwined:

When the tree is no longer approached merely as tree, but as evidence for [*Zuegnis für*: also evokes "testimony," "witness"] an other, as the location of *mana*, language expresses the contradiction that something is itself and at one and the same time other than itself, identical and not identical. Through the deity, language is transformed from tautology to language. The concept, which some would see as the sign-unit for whatever is comprised under it, has from the beginning been instead the product of dialectical thinking in which everything is always that which it is, only because it becomes that which it is not." (*DE*, 15)[16]

On this account, language is born with the recognition of a tension between identity and nonidentity. This recognition harbors both a sense of an excess, which simultaneously infuses and is beyond our experiences and concepts, and a sense that each being exceeds "itself" as a locus of dense, inexhaustible, and pregnant relations. With this sensibility, the possibility emerges of pursuing an elusive dialogical relation with the otherness that surrounds us. Beyond tautological closure, a journey of questions and partial, often oblique, and always fallible responses can be embarked on in relation to otherness. Here, cognition "does not consist in mere apprehension, classification, and calculation, but in the cancellation of each immediacy" (27). Enlightenment thus articulated is no longer the idea of a transparent illumination of all that is unthought but, rather, the illumination of an essentially inextinguishable region, beyond our horizon, that solicits an unending interrogative engagement with nonidentity.

Yet this dynamic and opening space of language *as such*—language that moves *away* from fearful and rageful tautological identifications and *toward* a greater sense of the reciprocal entwinement of differences, wonder, and an often agonistic appreciation—is stillborn. "This dialectic," write Adorno and Horkheimer, "remains impotent *to the extent that* it develops from the cry of terror which is the duplication, the tautology, of terror itself" (*DE*, 15–16, my emphasis). On their reading, *mana*, which indicates a realm of otherness, simultaneously establishes this realm as "the terrifying" and thus tends to preclude the interrogative relations that the acknowledgment of otherness might make possible: relations in which paralyzing fear might be diminished. In the structure of rigid taboos the identification of the terrifying fortifies itself against any encounter that might depressurize the situation of its genesis. Hence the duplication of terror.

There is good reason to believe that there is often far more wonder, understanding, and interrogative reverence than Adorno and Horkheimer recognize in "primitive societies" (a forced category). Yet their account is fashioned largely as part of a genealogy of modernity's darkness. They would illuminate by analogy modernity's own subversion of enlightenment, modernity's forgetfulness of the mythic (i.e., tautological) fear that lurks in its own positions. Thus they note in conclusion that *mana*'s establishment of an identity between the unknown and fear bears an uncanny resemblance with modern enlightenment's own relation to otherness. "Man imagines himself free from fear when there is no longer anything unknown. That determines the course of demythologization, of enlightenment. . . . Enlightenment is mythic fear turned radical. The pure immanence of positivism, its ultimate product, is no more so to speak than a . . . universal taboo. Nothing at all may remain outside, because the mere idea of outsideness is the very source of fear" (*DE*, 16).

But is this path of fear and mastery inexorable? Whenever humans live under conditions structured by overwhelming helplessness before nature and by antagonistic historical developments such as class subjugation, patriarchy, mindless instrumental reason, and culture-industry conformity, Adorno and Horkheimer think that fear and an insistent yearning to subjugate and extirpate will characterize our relation to otherness. They are also pessimistic concerning the prospects for substantial change in contemporary society. Yet they do not proclaim an ultimate *equivalence* between enlightenment and mythic closure, nor do they say that all enlightenment necessarily leads to self-destructive barbarism. They *distinguish* between (a) enlightenment as the possibility of language *as language*, which opens a negative dialectical relationship with the world, on the one hand, and (b) mythic tautology, taboo, and the devitalization of this dialectic, on the other. Historically, the two developments are entwined in ways that adversely affect negative dialectical relations. They emphasize the entwinement in order to highlight and resist the danger. Though enlightenment even under the best historical circumstances, can never rid itself entirely of its mythological dimension, it can and must continually "accommodate reflection on this recidivist element": that is to say, work to recall that which lies beyond its identifications and interrogatively engage it anew. It is this different and more desirable trajectory of thought—leading away from reciprocal enslavement and impoverishment, and toward dialogical coexistence with nonidentity—that glimmers as a hopeful possibility in the "earliest known stages of humanity."

Other traces of an alternative enlightenment are suggested in the chapter on Odysseus,[17] where the *Odyssey* is read as an allegory of the ways that "the identically persistent self which arises in the abrogation of sacrifice immediately becomes an unyielding, rigidified sacrificial ritual that man celebrates upon himself" (*DE*, 54). Thus, for example, in Odysseus's encounter with the sirens (who threaten his identity and the order to which he belongs), Adorno and Horkheimer see "a presentient allegory of the dialectic of enlightenment" and the subjugations entailed in a society based on class domination and the mastery of nature (34). Odysseus can survive the allure of what lies beyond his order only by being tied powerlessly to the mast ("just as later the burghers would deny themselves happiness all the more doggedly as it drew closer to them with the growth of their own power") and by plugging the ears of his oarsmen so that they labor in a disciplined, undistracted manner (34). "In class history, the enmity of the self to sacrifice implied a sacrifice of the self, inasmuch as it was paid for by the denial of nature in man for the sake of domination over non-human nature and other men. This very denial, the nucleus of all civilizing rationality, is the germ cell of a proliferating mythic irrationality: with the denial of nature in man not merely the *telos* of the outward control of nature but the *telos* of man's own life is distorted and befogged" (54).

To some, such passages are emblematic of Adorno and Horkheimer's hopelessly aporetical claim that rationality and selfhood is necessarily self-destructive.[18] To others it indicates that Adorno is unable to rid himself of a notion of a repressed "true self." I think both interpretations are wrong. When Adorno and Horkheimer write of a *telos* that has been distorted, they have in mind not a reified essence but the open-ended and agonistic (though not antagonistic) activity of interrogatively engaging nonidentity—nonidentity that has been subjugated by the constraints imposed by current identities. If we have a *telos*, it is only in the sense that we harbor this more promising possibility which leads away from totalizing forms of power and toward a "morality of thinking."

Hints of this alternative appear in the closing pages of the chapter on the *Odyssey*. "The innermost paradox of the epic," they write, "is that the concept of homeland [toward which Odysseus strives] is opposed to myth."[19] Transfiguring Novalis's assertion about the relation between philosophy and homesickness, Adorno and Horkheimer maintain that "all philosophy [that they could embrace] is homesickness," yet only if this is understood not as a yearning for a "primordial state of man" but as a longing for homeland, where "homeland is a state of having escaped."[20] Paradoxically, the self that gains a temporally expansive sense of purpose, coherence, and identity (in part by subordinating or sacrificing numerous possibilities—even in the best cases, when identity does not become identity-crazed madness), plays an integral and positive role in the endlessly unfolding drama of escape. For it is precisely such a self—constituted as a particular identity which holds itself together and resists dissolution and mergence with the impulses and demands of each moment—that can gain a self-conscious, temporally mediated distance between itself and events, through which a critical light can be shed on fatelike situations. And temporal mediation is crucial to the *Odyssey* insofar as "the translation of the myths into the novel [i.e., their Homeric late-epic elaboration] . . . pulls myth into time, revealing the abyss that separates myth from homeland and reconciliation." While this critical distance of a self that distances itself from the mythic immediacy of the present can in turn proliferate mythic subjugative relations and harden into a "cold distance" that separates itself from the world's suffering, "at the same time, [it] allows the horror as such to appear, which in [mythic] song is solemnly obfuscated as fate."[21] At a distance, horror *as such* can become visible insofar as it ceases to define existence exhaustively. It can appear as a distinct, finite foreground against the background of other possibilities.

The temporally extended self and narrative also engenders memory, which plays a role in a more hopeful enlightenment. Memory retains the experiences of subjugation, and can foster an ongoing discomfort that solicits our critical efforts. Further, insofar as the memory of a particular horror places the suffering in the past, it opens the hope and possibility that we might transcend suffering. "It is the self-consciousness in the moment of narration that stills terror. Discourse itself, speech in opposition to mythic song, the possibility of remembering the disaster that has occurred, is the law of Homeric escape. . . . Where the account comes to a halt, is the caesura, the transformation of the reported into something that happened long ago, and by virtue of the caesura the semblance of freedom lights up, which ever since civilization has not succeeded in extinguishing."[22] The idea of the caesura or break in the flow of narrative time is important because of Adorno and Horkheimer's sense of how the myth of uninterrupted temporal "progression" can itself partake in constituting a closed world.[23] It is with the engagement of a certain extremity of distance (viz., the cold description in the *Odyssey* of the hanging prostitutes, compared to birds caught in a net, captive bodies which "kicked their feet a little while, but not for long") that horror can be made to appear in a radical fashion which halts the flow and composure of the narrative (as we are weighed down by an unforgettable question: "Not for long?") and which places us in unwonted interrogative proximity to the world. This possibility of temporal distance, at once surpassing myth and checking the proliferation of the mythic in time itself, points toward an understanding of critical proximate distance that Adorno and Horkheimer view as integral to a "positive notion of enlightenment."

As with their discussion of *mana*, their normative project hinges on an understanding of "language as language," of "discourse itself," as essentially located on an edge between identity and nonidentity (entwined with temporal difference) which opens the possibility of a more generous and receptive dialogical engagement with others and otherness which exceeds the immediate present and which might mitigate suffering, expand freedom, and enrich our relations. If their account is paradoxical and never claims to have wholly escaped the mythical, this is owing not to a recognition of fundamental futility but, rather, to an awareness that reflection on these paradoxes and their accompanying dangers is inseparable from an enlightenment worthy of the name.

Adorno and Horkheimer's brief discussions of Judaism and Christianity also contribute to the positive project that builds energy in the margins of the *Dialectic of Enlightenment*. Indeed, their debt to these two traditions is far greater than the space they devote to them might suggest. Their engagements with these religions are replete with criticism of the subjugative aspects within each. Especially in Judaism, they see the pre-animistic terror in the face of nature "translated . . . into the notion of an absolute self which, as creator and overlord, completely subjugates nature." God as absolute sovereign is equivocal because he not only represents an overwhelming power but also "can liberate us from this cycle." Yet the absolute "*I am*, which tolerates no opposition, exceeds in its inescapable force the more blind, but therefore more equivocal assumption of an anonymous fate." This God is central to a subjugative relation in which all that is other is "entangled in the net of guilt" is and punished (*DE*, 177).

Nevertheless, entangled in this net of power, Adorno and Horkheimer draw our attention to developments that enhance the possibility of something quite different. In the prohibition on graven images and the ban on pronouncing the name of God, "Jewish religion allows no word that would alleviate the despair of all that is mortal. It associates hope only with the prohibition against calling on what is false as God, against invoking the finite as the infinite, lies as truth. The guarantee of salvation lies in the rejection of any belief that would replace it: it is knowledge obtained in the denunciation of illusion. . . . The justness of an image is preserved in the faithful pursuit of its determinate negation" (*DE*, 24). In calling us to this activity, this intransigent sense of nonidentical transcendence provides an ongoing check against the imperialism of identity which seeks to subsume the other within a closed totality. Properly rendered, the image/name ban again accents the centrality of positioning our understandings of language, freedom, and ethical relations on the *edge* between identity and nonidentity. By pronouncing the inadequacy of our fixed identities, the ban draws us to the edge where experience and language open beyond their extant formulations. Yet the ban simultaneously prohibits all claims to leap beyond the edge of our finitude to a transparent grasp of otherness. It highlights the dangers of complacency in both its finite and its infinite modes. By resolutely locating us at the paradoxical negative-dialectical limit, it thwarts mythic closure and enlightenment's tendency to revert to tautology.

The Jewish and Christian traditions also offer the concept of grace, present in Judaism in the relationship between God and the Chosen People and extended in Christianity (cf. *DE*, 177). However grace may function in the theological network of power, it also points beyond this entanglement as it symbolizes and calls us to the possibility of a generosity that might escape and overcome the "net of guilt and merit," the fateful accounting and equivalence of exchange that denies the nonidentical. While Adorno and Horkheimer's discussion of grace is brief—merely a passing comment—it is perhaps the central idea in the *Dialectic of Enlightenment*, which, by continually criticizing the myth of identical exchange (whether in the commodity form which reduces nonidentities to a uniform quantitative measure, or in totalizing concepts which claim to be identically exchangeable for the object, or in the representational substitution practices of magic, or in wage labor which, even at its best, calculates remunerative exchange in terms of the reduction of humans to quantities of labor), gestures in negative fashion toward an extravagant generosity that goes beyond the idea of *fair* exchange.

Adorno and Horkheimer push in this direction because of their sense that identity *always* does some violence to that for which it claims to be a perfect substitute. Thus, in a way, the very idea of "fair identical exchange" is oxymoronic. Or, better, read in light of the lie it bears within, identity points beyond itself toward an idea that would give the *nonidentical* its due (which, as we shall see, implies a concept of identity). But since this is always an incomplete project, giving (i.e., grace) must strive to go beyond the calculated exchange that is recognized as due. This is evoked when the idea of grace is disentangled from the subjugative aspects of monotheistic metaphysics and is read as a generosity that seeks to recognize the infinite possibilities of the other which lie beyond the calculating grasp of oneself or even of the other. In this light, the very being of language *as language*, of time *as time*, "in which everything is always that which it is, only because it becomes that which it is not" (*DE*, 15), emerges from and as a grace that exceeds the identity-driven "net of quilt and merit" and dialogically extends toward the possibilities of otherness as yet unrecognized.

Grace is vital to the ethic that animates the *Dialectic of Enlightenment* in another related sense. One is not simply the *subject* of identifying violence but also the *object*; and not simply the object of oneself, but also of others. Moreover, we undergo arbitrary suffering from nonhuman nature. To the extent that the identity-saturated tropes of guilt, punishment, and mastery govern our response to this situation, Adorno and Horkheimer maintain that we "reduplicate" the terror we face and increasingly subjugate and impoverish both the other and ourselves in a vicious cycle. The movement beyond totalizing identity requires relinquishing—or at least loosening—the identity-driven will to respond to violence and threat according to the logic of an eye for an eye. Grace as forgiveness (and a certain kind of forgetfulness?) is requisite for grace as receptive generosity.

From this discussion, a "positive notion of enlightenment" begins to emerge. Yet even if directions are suggested other than the dark paths that the *Dialectic of Enlightenment* tenaciously traces with its genealogies, is there any reason to hope that we might develop those insights in ways more likely to flourish and less prone to succumb to the dialectic of enlightenment? And without God and *mana*, can such directions be cast in terms sufficiently desirable and compelling? What does negative dialectics draw on for guidance and sustenance, vision and energy? These questions are central to the rest of the present chapter.

As mentioned, Adorno did not live to write his work on morality. In a sense, though, Adorno was always already writing his moral work, whatever else he was writing. I don't mean that Adorno blindly collapsed the diverse concerns of the world into a one-dimensional moralism, but rather, as I noted above, that he shares a certain proximity to Levinas with his sense that thought is rooted in a finite and indeterminate opening toward nonidentity that is most fundamentally ethical.[24] Insofar as knowledge and art ceaselessly spring from the giving and gift of this fragile opening, they must always gather around it, practice it, describe it, protect it, speak of the violence that threatens it and, most important, solicit our renewed efforts to reopen this tragic (but least tragic) possibility. Where there is thinking, there is a "morality of thinking"; where there is art, a morality of art. Adorno's *prima philosophia* is a nonidentical first, an active-passive moving, lingering, slipping away toward the elusive otherness of the world. I am not sure what Adorno's work on morality would have looked like. But perhaps its absence is not so important for the philosopher who wrote that "the crux of a philosophy is what happens in it, not a thesis or a position— its texture" (*ND*, 33). It is difficult to imagine that what might have happened in the moral work would have been significantly different from what happens in the works we have; and the moral declarations would certainly have revolved around that happening. So let us turn to the texture of Adorno's thought.

The Morality of Thinking and the Gift

Adorno's morality of thinking develops themes, practices, and solic-itations toward a dialogical receptive generosity. Yet there is no thesis that cannot be assimilated or that will not drift off to sleep in ways that legitimate, conceal, and proliferate ignorance and subjugation. Sensing this, Adorno places his hope in constellations of agonistic ideas that depict and solicit not simply particular thoughts (concerning receptive generosity, nonidentity, and so forth) but, more important, a *style of thinking* animated by concerns that demand a difficult receptive en-gagement with otherness. For Adorno the truth of such combinations lies not only in the thoughts conveyed and juxtaposed, but also in the extent to which they engender an activity of thinking that embodies and enlivens the constellation's tension-laden themes. We shall miss nearly everything unless we're attentive to the ways in which what is written works to solicit a morality of thinking.[25] Vital in this regard is Adorno's thought that "instead of reducing philosophy to categories . . . [philosophy's] course must be a ceaseless self-renewal, by its own strength as well as in friction with whatever standard it may have" (*ND*, 13). Philosophy—indeed, thinking—is not only *about* the nonidentical but is a *relationship with* the nonidentical. Hence, it must neither come to rest in itself nor allow the serious reader to do the same. To under-stand Adorno's thoughts here and gain a sense of his practice, we turn to the web of insights that he weaves around the questions of identity and nonidentity.

Negative Dialectics is animated by many paradoxes, but none are more central than those which concern the relation between identity and nonidentity. The text lives, breathes, and feels its way about on the frontier between its sense of "the untruth of identity, the fact that the concept does not exhaust the thing conceived," and its knowledge that "the appearance of identity is inherent in thought itself, in its pure form. To think is to identify" (*ND*, 5). Whenever we think, the very conceptuality through which we strive to comprehend the world engenders concealment. The universal character of concepts always says more (by importing connotations) and less (through abstraction) than the particularities of the world toward which concepts aim. The world that surrounds and includes us, Adorno claims, is irreducibly nonidentical, persistently exceeding our grasp. Conceptual movement toward the world continually falls short owing to the concept's forgetful claim to be exhaustive of its object. In the worst cases, this forgetfulness takes its own "demand for totality" as "its measure for whatever is not identical with it," systematically excluding the nonidentical within a totalizing conceptual order (5–6). Even in negative dialectics, however, thought must endlessly struggle against its forgetfulness in order to reflect on its violence—not only engaging the protean specificity of the world, but letting it speak as well.[26]

Thus Adorno defines his project: "The name of dialectics says no more, to begin with, than that objects do not go into their concepts without leaving a remainder. . . . Dialectics is the consistent awareness [*Bewusstsein*: also means "consciousness"] of nonidentity" (*ND*, 5 trans. altered).[27] This observation stems from a retrospective sense of the eclipsing and damaging aspect of concepts: "My thought is driven to [an awareness of nonidentity] by its own inevitable insufficiency, by my guilt of what I am thinking" (5).

I pause here, for questions immediately arise concerning the status and philosophical underpinnings of such thoughts. Perhaps no one has probed this terrain more provocatively than Albrecht Wellmer.[28] I turn to Wellmer's Wittgensteinian-Habermasian critique because Adorno's possible response helps one clarify and appreciate the insights of negative dialectics.

Wellmer's "metacritique" hinges upon two key claims. First, Adorno's description of the "rigidity" and fixed monotonous generality of concepts, which is central to his "totalizing" critique of the violence of concepts as such, remains tied to the "rationalistic fiction" (that words ought to represent things in a singular manner) from which it seeks critically to distance itself. In contrast, writes Wellmer, Wittgenstein illustrates that "words can be used in many and various ways" and that their character is better evoked by "the image of family resemblance, and also, that of the rope that consists of a multiplicity of fibers. . . . [T]his multiplicity of ways of using a word reflects the openness of linguistic meanings."[29] As multiplicitous, flexible, and open-ended, language *as such* is hardly violent. Rather, only *particular uses* of concepts can be thus depicted: "specific disturbances, blockages, or limitations of communication" that stand out against the intralinguistic normative backdrop of unimpeded communicative practice.[30] Second, accompanying Adorno's "rationalistic fiction" is a "residue of naivete" through which Adorno adopts a position *outside* the linguistic realm in order to condemn the latter's relation *as such* to the extralinguistic. From no other perspective could Adorno assert the injustice of "*the* identificatory concept."[31]

Considering the first point, Wellmer wrongly attributes a "rationalistic fiction" theory of concepts to Adorno. Admittedly, it is possible to rely on a position one repeatedly criticizes. But the depth of Adorno's criticism is expressed most significantly by the fact that he always participates in and, in the affirmative aspects of his work, thinks about the world of language from a perspective which is *different from* the rigid, rationalist theory he criticizes. The multiplicity, flexibility, and openness of concepts is integral to Adorno's understanding of language, as is clear in his discussion of "constellations." Adorno writes: "The model for this is the conduct of language. Language offers no mere system of signs for cognitive functions. Where it appears essentially as language, where it becomes a form of representation, it will not define its concepts. It lends objectivity to them by the relation into which it puts the concepts, centered about a thing. Language thus serves the intention of the concept to express completely what it means" (*ND*, 162).[32] A concept expresses different meanings, depending on the other concepts in a constellation. Different constellations, writes Adorno, change the categories within them, and "when a category changes . . . a change occurs in the constellation of all categories, and thus again in each one" (166). Hence, *contra* Wellmer, Adorno (by other paths) arrives at a view of language "as language" that shares with the Wittgensteinian paradigm a sense of flexibility, multiplicity, and openness. This understanding is also articulated in the *Dialectic of Enlightenment* where language emerges *as such*, beyond tautology, when each concept, as Adorno says elsewhere, "is always that which it is only because it becomes that which it is not."[33]

Despite this subtle understanding, Adorno attributes to concepts and language *as such* an unshakably blind and transgressive quality. Yet he does so not from a position outside language but, rather, rooted in the knowledge that "there is no peeping out" and that there is "no way but to break immanently" (*ND*, 140). Thus it is important that immediately following the passage on dialectics as the "awareness" of nonidentity, Adorno adds that dialectics "does not begin by taking a standpoint." It originates not from a stable, extralinguistic ground but from his "own inevitable insufficiency," from a personal "guilt of what I am thinking" (5). The awareness of nonidentity emerges as a *personal* and *privative* sense, not as an impersonal positive consciousness. The privation is made conscious *immanently* and *retrospectively* as one endlessly discovers particularities and possibilities that one (or others) previously concealed and transgressed. The essentially immanent and performative aspect of this awareness is a main reason why Adorno warns in part two of *Negative Dialectics* against making nonidentity just another ontology. One does not "peep out" and see with a transparent affirmation; rather, one is driven to an "awareness" (expressing a sense that is more experiential, less completely determinate, positive, or singular than a "standpoint") of the partly transgressive aspect of one's relationship to nonidentity. This awareness is a negative sensibility that emerges as one—driven in part by repeated guilt—reflects on one's own connection to that which appears repeatedly as a *more-having-been-taken-for-less*.

Certainly the "more" always appears within the linguistic realm: "What would lie beyond makes its appearance only in the materials and categories within" (*ND*, 140). Yet this fact need not limit us to an understanding of identity and difference (or the "more") as something that concerns only the linguistic realm. Instead, through our retrospective reflections on the limits of every concept and every constellation of concepts, we can become aware of an extralinguistic surplus that is tragically eclipsed (though no particular eclipse is positively beyond our limits to rescind). In this general sense, that which lies beyond and which appears obliquely and privatively "within" is the existence of a "more" that is always partly damaged by linguistic thought. The importance of the *privative* character of this knowledge is indicated in

Adorno's defense of relativism (of which he was otherwise very critical) against critics who accuse it of assuming "one absolute, its own validity." These critics "confuse the general denial of a principle with the denial's own elevation to affirmative rank, regardless of the specific difference in positional value of both" (35–36). Similarly, Wellmer confuses Adorno's position, rooted in a sense of limits and in a general denial of harmonious identity, with an affirmative position rooted in a positive and absolute view from without. The difference here is between a recognition of inexhaustible extralinguistic horizons versus a determinate claim to know an object.

One might grant here that Adorno's view of language is subtle, and avoids reliance on a naive extralinguistic position. One might even concur that language always involves concealment. But what of the claim that concealment is essentially damaging and unjust? Certainly this is true of a lot of language, but we can all think of linguistic expressions for which—no matter how incomplete—claims about damage and injustice seem out of place, unless we stretch these terms beyond recognition. How, then, are we to understand Adorno?

I think we move closer to the sense of Adorno's position when we consider language not simply as a theoretical illumination or revealing (and concealing) of the world, but as essentially entwined with ongoing and active practical relations with the world. Thus, as we move and act in the world, that which is concealed is highly liable to various kinds of damage: beings we step on inadvertently, beings whose manifold particularities and possibilities we preclude and transgress, beings with no place in the orders of practice that we construct and fortify. When conceptual thought is considered as embodied in practical relations, claims about the pervasiveness of damage and injustice become more plausible.

If we read these claims as universal ontology, however, they still seem to stretch the truth beyond the breaking point. There are many concealments—e.g., those which are highly temporary in nature, or of seemingly marginal significance from almost any imaginable perspective, or so minor compared to what they help reveal—that it makes little sense to call unjust and transgressive.

Adorno, then, if he is offering a universal ontology here, is pushing too far. Yet I suggest that we read his claim not as ontology but as part of a project aimed at ethically opening practical engagements with otherness. From this vantage point, the claim is not that every single concealment is unjust and violent, but that all people live lives resulting in a substantial amount of blind and unjust transgression. To reflect on the human condition is to become aware, in a general sense, of our implication in a vast wake of specific damages of which we are at any moment largely unaware. This is the ontological claim. Given this retrospective on our condition, Adorno projects an exaggerated claim about the relation between concealment and violence, not as an ontology but as part of an ethical constellation composed in such a way that we might think against the grain of the propensities for violence lodged in our perception and thought. Not all unconscious concealment is violent, but much of it is; and all of it which is, is so prior to our awareness. Hence, for ethical reasons, Adorno projects a universal *suspicion* of violence in the form of exaggerated claims. From an ethical vantage point, such exaggeration has truth, insofar as anything less would allow us too easily to say, "There is no damage *here!*" This understanding of Adorno makes the most sense to me, both in terms of the subject matter and with respect to his overall project. It is, at any rate, the understanding that infuses what follows.

Adorno articulates his sense of nonidentity as an extralinguistic realm that is "always already" and "not yet" with respect to language. The nonidentical is "always already" because our linguistic consciousness emerges from and is colored by a dense corporeality having multiplicitous material relations with the world.[34] Adorno writes: "The somatic moment as the not purely cognitive part of cognition is irreducible. . . . Physicality emerges . . . as the core of that cognition," displacing the dream of a conscious or intersubjective constitution of the body (*ND*, 193–94). The nonidentical is "not yet" insofar as linguistic thoughts "point to entities" that they have not posited, entities whose fundamental nonidentity cannot "be abolished by any further thought process" (135–36). Far from the sovereign self-giving subjectivity exemplified by Kant, we are always already somatically receiving the extralinguistic, and thoughts are engaged with a nonidentical world that they must strive to receive.

Since "subject" and "object" reciprocally permeate each other, their fates are likewise entwined (*ND*, 139), and care for the self and care for others and otherness are ultimately inseparable. As the *Dialectic of Enlightenment* powerfully illustrates, the effort to master the otherness with which one is entwined blinds one to the world and to oneself, and ends in a sacrifice of the freedom, vitality, and wealth of the self and

otherness. This is not to imply that receptive generosity toward nonidentity, for Adorno, ultimately boils down to a concern for the self—to
an enlightened self-interest in subtle negative-dialectical garb. Rather, a
sense of the entwinement of identity and nonidentity leads him to the
position that there is no identical self standing beyond nonidentity that
could calculate the utility of otherness. Instead, the self *is* through this
generous and receptive relation to otherness: a relation that always exceeds one's calculations and consists in the movement *beyond* these
identities, beyond the "transgression of self-enclosures,"[35] however
much they are necessary for this movement. And the "happening," the
"performance," of negative dialectics, hinges on a receptive generosity
toward nonidentity in the *first* moment (and throughout). Again, not as
calculation but as the very questioning of what is valuable and as the
very origin of value in this questioning. If a fateful equivalence governs
the parodic reciprocity of subjugation and impoverishment, the nonidentical reciprocity of freedom, wealth, and ethics is born in a generosity that seeks—as grace—to exceed equivalence.

I have outlined Habermas's equation of Adorno's sense of tragic nonidentity with nihilistic despair. Yet, in contrast, for Adorno, our awareness of tragic finitude is essential for beginning to address and move
beyond a hubris (institutional, epistemic, personal) that otherwise proliferates a blindness which enslaves and devitalizes the self and the surrounding world in a parody of reciprocity. With a tragic sensibility and
a sense of grace, it is possible to demythologize forms of subjugation,
illuminate concealed suffering, and carefully extend oneself, as both
imperfect witness and as imperfect participant, toward fragments of the
nonidentical world's protean "color."

Crucial to this possibility is Adorno's claim that "the force of consciousness extends to the delusion of consciousness" (specific delusions
and thought's general awareness of its ineliminable delusive aspect), for
if this were not so, it would appear that all efforts to think would be a
priori equally blind (ND, 148). The substance of Adorno's claim lies in
his idea that the will to identity (which when unreflectively unleashed
pushes toward a seamless transparent totality) is capable of discerning,
when cautiously and self-critically deployed, the presence within its orders of that which defies logical transparency. It is, in other words,
capable of discerning the absence of its own ideal; and this inadequacy
is pressed into the form of a contradiction whenever thought loses
sight of it and sinks into an exaggerated sense of itself. Contradiction is

a persistent feature of self-inflated thought, insofar as its efforts to neatly subsume the world lead it into a process of deceitful conceal- ment which increasingly belies the claim to identity that spawned this process to begin with. Thought's shortcomings here harden into denial with respect to nonidentity. Yet the antagonism between the claim to truth and the demands of concealment give rise, if not to a necessity, at least to a very strong propensity to generate contradictions internal to thought. Thus Adorno writes: "The antithesis of thought to whatever is heterogeneous to thought is reproduced in thought itself, as its im- manent contradiction" (146). As long as the subject does not exhaust the world around it and cannot entirely thwart existential pressures to conceal the finitude of its claims, our thought and experience will har- bor contradictions that we are capable of critically discerning. Such a capacity, which is a condition for an awareness of tragic finitude and the more desirable directions of negative dialectics, is in turn enhanced by cultivating that awareness.[36]

Developing the more hopeful capacity of negative dialectics to "in- terpret every image as writing" and "show how the admission of falsity is written in the lines of its features" (*DE*, 24), Adorno focuses on dis- cerning contradictions between different layers of abstraction and con- creteness, in a "reciprocal criticism of the universal and particular . . . whether the concept does justice to what it covers, and whether the particular fulfills its concepts" (*ND*, 146). Through a reciprocal critique between universals and particulars (not unlike Kant's play between the imagination and reason, but with a greater tragic sensibility), Adorno seeks both to expand and improve our grasp of each so that they might do more "justice" to the world, and also to distance us from the hold each has on us. The expansion of thought and a cultivation of critical distance are both integral to a dialogical freedom that emerges in thought's "revolt against being importuned to bow to every immediate thing" (*ND*, 19).

Thus partially released from the often spellbinding power of a con- cept, image, or object, Adorno writes that the self cannot only pursue immanent critique, but can also partially escape dialectical immanence itself. It can "withdraw to itself, and to the abundance of its ways to react" (*ND*, 31). This greater degree of distance and spontaneity is vital to the subject's effort to improve its critical engagement with the world. Yet Adorno solicits further dialectical self-reflection in order to check the self's tendency to mythologize its more spontaneously generated thoughts into another absolute. Just as the power of thinking lies in the discrepancy and reciprocal criticism between the universal and the particular, Adorno emphasizes the capacity to slip to and fro between an immanent perspective and a wilder, more external spontaneity:

> Mobility is of the essence of consciousness; it is no accidental feature. It means a doubled mode of conduct: an inner one, the immanent process which is the properly dialectical one, and a free, unbounded one like a stepping out of dialectics. Yet the two are not merely disparate. The unregimented thought has an elective affinity to dialectics which as criticism of the system recalls what would be outside the system; and the force that liberates the dialectical movement in cognition is the very same that rebels against the system. Both attitudes of consciousness are linked by criticizing one another, not by compromising. (31)

Through the constellation of agonistic critiques of identity and nonidentity, universal and particular, immanence and "stepping out," a proximity and a distance are established between self and otherness; and, thereby, the greatest possibility emerges for practicing receptive generosity.

Thus we see elements of Adorno's effort to move in directions less exclusive, more fluid, and pregnant with possibility. To say that our inability to break completely free of blinding finitude renders negative dialectics futile is true only if we use a standard of total transparency. Otherwise, that "failure" simply means that our paths are never complete and never guaranteed. Nothing absolutely precludes the possibility that one day (or, perhaps worse, never) we may discover blindnesses in our own thinking and practices that are far more dangerous than those against which we energetically struggle today. Nevertheless, Adorno argues, our best chance for lessening such suffering and blindness, as well as our richest possibilities, lie in negative dialectical engagements between identity and nonidentity. Adorno esteems this transfiguring movement of thought, not simply as a means to the end of recognizing otherness but as a *way of being* that is itself the highest: namely, the practice and texture of freedom, wealth, and receptive generosity.

Adorno further articulates the ethics of negative dialectics by drawing on the idea of "identification [as] reflected in the linguistic use of the word outside of logic, in which we speak, not of identifying an object, but of *identifying with* people and things" (*ND*, 150, my emphasis). Adorno refers not to a homogenized sense of belonging with others but, rather, to "a togetherness of diversity" (150). If "the copula says: It is so and not otherwise" (147) and, in so doing, ideologically subsumes a world not predesigned for our categories, negative dialectics takes nonidentity to be "the secret *telos* of identification" (149). This statement expresses more than a yearning to affirmatively comprehend otherness in a manner that leaves the self and the other unchanged (except for the newfound recognition). Of course, the event of togetherness as a greater mutual recognition of diversity is itself grand to Adorno. Yet he also suggests a more dynamic sense in which "togetherness of diversity" is a mutual transfiguration—entwined with, but not limited to, enhanced recognition of extant otherness—that lessens senseless suffering and partakes in opening and elaborating diverse possibilities for greater freedom and qualitative richness. Adorno writes of an activity that seeks illumination through a mutual inscribing which, "in placing its mark on the object . . . seeks to be marked by the object" (149). With this idea of mutually transfiguring impressions, he suggests transgressing self-enclosures and a mutual, often agonistic writing of richer possibilities.

"Identifying with" is an enormous task, and calls for a close reading. The difficulty is connected with the blindnesses of identification, and with Adorno's far-reaching—if not ontological—claim that, as expressed in the last line of the *Dialectic of Enlightenment*, "all things that live are subject to constraint" (258). As such, beings collide repeatedly with the thwarting boundaries of their situation; scars often form, fostering rigid self-protective limitations and horizons. Moreover, Adorno argues, human history has to a large extent intensified antagonisms that mold, permeate, and subjugate people, so that the contradiction between subject and object is not most basically "a cogitative law." Rather, "it is real." The agony of negative dialectics, then, "is the world's agony raised to a concept" (*ND*, 6).

Thus, "identifying with" cannot have mutual recognition of extant diversity as its sole or deepest aim, though recognition of diversity is important for the further dialogical engagement and togetherness that Adorno suggests. Even if a certain acceptance of multiple scars might be vitally important for the coexistence of diversity, a worthy mode of the latter would certainly tend in part toward more than this—lest it understand the recognition of nonidentity as culminating only in an embrace of other subjugated and subjugative identities. This insight deeply informs Adorno's understanding of mutual inscribing and agonistic cooperation.

Adorno seeks to transfigure not simply our own identifications but objectification more generally, which emanates from multiplicitous sources (including the "object" itself when it is another self). Thus he writes of "thought forms [that] tend beyond that which merely exists," not in order to control otherness but to "heed a potential that waits in the object" or in the other that has been suppressed or undeveloped. He seeks a "resistance of thought to mere things in being, [a] commanding freedom of the subject, [that] intends in the object even that of which the object was deprived by objectification" (*ND*, 19). Adorno sometimes formulates this position in terms of the relation between labor and the world, while at other times he uses dialogical tropes. Consistently, however, he describes the conditions for thought's movement beyond the extant by juxtaposing radical receptivity with strong activity. "Our aim," he writes, "is total self-relinquishment" (13). Similarly: "If the thought really yielded to the objects . . . the very objects would start talking under the lingering eye" (27–28). Yet the telos of this "yielding" includes possibilities still unacknowledged and unexplored in, or by, the object (or other). Hence Adorno agonistically juxtaposes to the language of yielding that of "a commanding freedom" which aims at transfiguring otherness even as it yields to it. Given the many points that proliferate subjugation, "to give the object its due instead of being content with the false copy, the subject would have to resist the average value of such objectivity and to free itself as a subject" (170–71). If the object is a subject, this response often involves resisting the other's own self-descriptions. While the self attempts to "give the other its due," freeing otherness from mythic identifications, this offered freedom in turn demands the freedom of the self, since "the object . . . only opens itself to the subjective surplus in the thought" (205). For Adorno, freedom and richness demand that both subject and object (where both are selves) partake in a reciprocally transfiguring giving and receiving.

In this vein, Adorno writes that the substance of a "changed philosophy" would "lie in the diversity of objects which impinge upon it and of the objects it seeks. . . . [T]o these objects, philosophy would truly *give itself*, rather than use them as a mirror in which to reread itself"

(*ND*, 13, my emphasis). This infinite task of giving must articulate itself in proximity with a radical vision of receiving that aims at "nothing but full, unreduced experience" of the other (13), one that "means to do justice to the object's qualitative moments" (43), including those unrecognized by the other. This effort to give by means of a radical receptivity and to receive by means of a radical generosity is wrought in the midst of a tragic finitude which involves risks, assimilation, and blindness. Yet this is the texture of a negative dialectics from which we have much to hope; and—in all its agonistic complexity—this is what it means "to love things" (191).

Generosity (cultivating both the gift beyond equivalence and the truncation of revenge symbolized by grace) is also central to Adorno's *Aesthetic Theory*, where he emphasizes repeatedly that the unity and form of an artwork, if it is not to suffocate itself, "must give the heterogeneous its due." Adorno writes of "a certain generosity or plenitude to art works which is diametrically opposed to the discipline inherent in them" (*AT*, 273). Moreover, "a posture of generosity" manifesting itself in the texture of an artwork (seeking to transcend past blindness and expose the violence of its own form) is an "essential aspect of accomplishment" (269). And giving and receiving finds no stronger expression in Adorno's work than in *Minima Moralia*, in a section entitled "Articles may not be exchanged," where he writes: "Every undistorted relationship, perhaps indeed the conciliation that is part of organic life itself, is a gift" (*MM*, 43).

Amid a melancholy dirge proclaiming that "we are forgetting how to give" (*schenken*),[37] Adorno hints at the forgotten practice. "Real giving," he writes, "had its joy in imagining the joy of the receiver. It means . . . going out of one's way, thinking of the other as a subject." This "going out of one's way" deviates from the way one goes about oblivious to others. Instead, one chooses to extend oneself—and be extended— great distances toward other perspectives and other joys, to entwine one's joy and thought with joys and thoughts which remain nonidentical even as they are engaged. Real giving, like negative dialectics itself, is about a change in direction; and only in the activity of giving are the powers essential to freedom and "life that lives" nurtured (playing on Kürnberger's line at the beginning of *Minima Moralia*: "Life does not live" (19). In its absence "wither the irreplaceable faculties which cannot flourish in the isolated cell of pure inwardness, but only live in contact with the warmth of things." Noting the entwinement of obliv-

ious conceit and self-subjugation, Adorno writes that those who become incapable of giving "freeze" (*MM*, 42–43). This passage should
not be read as an affirmation of giving that hinges on benefits for "the
self." It is rather a transfigured construal of the very genesis of self in
receptive generosity.

Giving entails an ability to receive the other's specificity, and this
receptivity is entwined with an ability to extend ourselves generously
toward others and otherness. Yet, as we have seen, this circularity cannot revolve around any simple affirmation of extant particularity. When
Adorno writes of giving in light of our receptive engagement with the
other's different perspectives, thoughts, desires, and joys, he doesn't
mean that we should strive simply to mirror others as they currently
are and give them simply what they desire. (As few illustrate so poig-
nantly as Adorno, for many people in contemporary capitalism this
might mean primarily that we ought to give more soap operas, plastic
smiles, Big Macs, "gift articles," exchangeable gifts, or some other
more "pseudo-individualized" item coughed up by the commodity culture.) Yet neither does Adorno mean that "real giving" is based on an
undistorted grasp of the other's essence, for he repeatedly submits the
ideas of undistorted thought and essence to incisive criticism, arguing
that all thinking involves blindness and that human beings are thoroughly permeated by history's contingencies and violence.

But then what *does* Adorno mean by "real giving," by "undistorted
relationships" that give subjects their due? His response refers, paradoxically, to the very *activity* of giving and receiving. If "every undistorted relationship is a gift," if life, freedom, well-being, and ethics
"freeze" in the absence of relations of giving and receiving, then the
giving of what is due to another self is precisely that which draws the
other to engage in the transfigurations of giving and receiving through
which "life can live." Essential to the gift is the solicitation to pursue
life through an ethic and activity of receptive generosity with otherness. Hence, giving must navigate the tensions between receptively addressing the other's extant perspectives, desires, and joys, on the one
hand, and responding to them in ways that might enhance the other's
capacity to receptively and generously engage the world, on the other.
Ignoring the former imperative leads to a blind imperialism; ignoring
the latter leads to a slackening of the will to resist and move beyond the
life-stunting limits of present beings. These two imperatives coexist in
mutual interrogation. Situated in their strife, we are most capable of a

gift that draws the other toward receptive generosity. Adorno is distressed by the decay in giving symbolized by the attitude of "take this, it's all yours, do what you like with it; if you don't want it, that's all the same to me . . ." because such an attitude is entwined with a "withering of irreplaceable faculties" on the part of the giver. Equally important, though, this attitude drains the gift received by the other of the tension-laden distance which is an ineliminable aspect of a real gift. A gift should not collapse to the perspective of the other, for that would imply that the ideal gift would "allow the receiver to give himself [the] present, which is . . . in absolute contradiction to the gift" (*MM*, 42–43). Even as the giver must extend herself toward the other, she also establishes in the gift a distance that draws the receiver beyond herself as she engages the gift. Thus, the gift is *not* "all yours": it must *not* eclipse its heteronomous quality with "do what you like" and "all the same to me." The gift engages the other in an agonistic dialogue. Given the tragic erring that accompanies human beings, we must move with caution and humility, but we should not seek to substitute a totalizing other for the totalizing self.

Some of Adorno's reflections on "possibility" and "color" further develop these directions. Adorno writes that art and critical theory must be animated by "the resistance of an eye that does not want the colors of the world to fade" (*ND*, 405). A significant part of this resistant yearning aims to grasp aspects of the other through constellations of insights that, with their juxtapositions and tensions, "illuminate the specific side of the object, the side which to a classifying procedure is either a matter of indifference or a burden" (162). "Things," Adorno writes, "congeal as fragments of that which was subjugated; to rescue it means to love things" (191). Adorno evokes a solidarity with the specific, an awareness that freedom and well-being emerge only through a generous receptivity that draws close to the particularities which lie beyond us and which are continually "fading" in the face of insistent identities, abstracting universals, and structures of power.

But, somewhat paradoxically, Adorno's sense of not wanting the colors of the world to fade extends beyond this proximate solidarity with specifics. An essential aspect of the world's colors, as multiple and distinct (in opposition to an anonymous hue in which we are simply submerged), is their pregnant being-at-a-distance. Color in this sense is that which—even as it appears through a constellation as vibrant and proximate—retains a distance, opens onto even more that is not yet grasped. When all sense of distance is lost, color rapidly becomes an indistinct anonymous surrounding. To "not want the colors of the world to fade" is to seek to preserve this sense of distance even as one draws near to the other. And it is the soliciting of this distance (not simply as a moment of the self's relation to otherness, but as a moment within otherness itself) that prevents the other from hardening into the seamless totality of any specific thing. The affirmation of the essential distance of color overflows the relation between self and other, flooding into the self's understanding of the other's relation to itself. Soliciting this sense of distance and engaging its possibilities is a significant aim of the gift.

Adorno expresses all of this in a dense passage: "Dialectics inclines to content [read: specificity] because content is not closed." He offers further articulation:

> To want substance in cognition is to want a utopia. It is this consciousness of possibility that sticks to the concrete, as the undisfigured. It is the possible, never immediate reality which blocks off utopia; this is why it [utopia] seems abstract in the midst of extant things. The inextinguishable color comes from nonbeing. Thought is its servant, a piece of existence extending—however negatively—to that which it is not. The utmost distance alone would be proximity; philosophy is the prism in which its color is caught. (*ND*, 56–57, trans. altered)[38]

The desire for substance and color expressed here must always articulate itself through movement toward specificities that are nonidentical to it; if it did not, it would go blind in the grayness of its totalizing identities. Yet we see that, in and through this yearning for nonidentical substance, there is also—perhaps even foremost—a yearning for *possibility*, for opening as such, that is an essential aspect of the desire for the concrete as something different and distanced from the disfigurations of identity. "Content is not closed," for its elusive presence is always partially beyond thought and is thus emblematic of an inexhaustible "possibility" that always "sticks" to thought: the possibility of more color and less senseless damage. Substance provides a site of possibility and solicitation into which thought might extend itself and undergo transfiguration. The claim isn't that the concrete *is* in itself undisfigured, but that it stands for, is *as*, that which remains at a distance from the disfigurations of identity. It thus evokes the idea of utopia as the coexistence of diverse beings beyond hostile antagonism. This soliciting idea is stifled by "the possible" (*das Mögliche*) insofar as the latter, as ideologically defined, predetermines and closes the world. "The possible" (and the "not possible") is that by which an order bestows on itself an aura of immutability, in the face of which a more open and indeterminate "possibility" withers.

Yet, recalling the last words of the *Dialectic of Enlightenment*—"all things that live are subject to constraint"—we can begin to understand how "utopia seems abstract in the midst of extant beings," even if it lives in the idea of nonidentical substance. For we are all both objects and proliferators of the closure of possibility. Hence the animating paradox of the present passage. On the one hand, negative dialectics must extend itself receptively toward nonidentical substance and specificity in an effort to interrogate critically the grayness of its own identities in the light of the colors emanating from the particularities and possibilities beyond its horizons. On the other hand, substance (others in their specificity) is not simply color but is also the object and bearer of color-extinguishing antagonistic coercion. Things (including human beings) are not wholly what we should love, but are "the marred figures of what we should love" (*ND*, 191). As such, they solicit our efforts to employ a sense of distance, *which they offer us* in their nonidentity, in order to offer *them* a sense of nonidentity and possibility with respect to themselves and the world. Through a painstaking giving and receiving that is never wholly devoid of damage and the possibility of blundering, thought is the "servant" of nonbeing, "a piece of existence extending—however negatively—to that which it is not." Such philosophical activity, which would draw close and seek to "live in contact with the warmth of things" through an engaging recognition of distance and nonidentity, is the prism in which our best color is received and given.

It is apparent in this discussion that, in opposition to antagonism or homogeneity, Adorno offers something more subtle, complex, and interesting than an ideal of static coexistence of diverse undistorted beings placidly observing one another. He locates an ethical relation in a dialogical generosity and grace articulated through the difficult relations which are the condition for freedom among interdependent nonidentical beings. Such an appreciation and generosity has a significant agonistic dimension. Through mutual encroachment, it seeks not to dominate but to critically reveal subjugations and explore eclipsed possibilities of thriving. In short, Adorno seeks to move beyond antagonism toward the agonisms of generosity. It is this agonism, not some sort of essentialism, I think, that most fully expresses the meaning of his reflection that "every undistorted relationship, perhaps indeed the conciliation that is part of organic life itself, is a gift."

Ethics as a Constellation of Solicitations

This interpretation of Adorno's morality of thinking as agonistic receptive generosity raises important questions. Adorno's texts contain passages that seem to evoke ideas of total reconciliation, total peace, and total transparency as the *telos* of negative dialectics. What sense can my interpretation of Adorno make of such passages? Furthermore, insofar as "conceptual order is content to screen what thinking seeks to comprehend" (*ND*, 5), how is one to write in a way that evokes, exemplifies, and solicits an ethical activity which appears to be in direct opposition to these very somnambulistic tendencies (not to mention many organized forms of social power)?

These two questions must be understood in relation to each other. Adorno's employment of concepts of total reconciliation must be read as part of his response to the difficulties both of thinking and of soliciting thinking; and his response to those difficulties cannot be understood unless one grasps the meaning and function of his paradoxical juxtaposition of total reconciliation with powerful evocations of such reconciliation's impossibility. Central to Adorno's approach to both questions is his idea of "constellations," which I shall now explore more fully.

I have argued that Adorno's disclosure of the blindness clinging to concepts leads him to embrace philosophy as an *activity*, a happening, that is not exhausted in a set of declarations and involves a performative dimension of continual striving beyond itself toward eclipsed otherness. Constellations are vital to this project in three ways. First of all, they push the task of interpretation beyond the idea that a single declaration or sum of declarations can grasp an object of thought. Interpretation now becomes a manifold process in which each declaration or concept is not only *supplemented* by others but also acquires its meaning precisely *through* "the relation into which [the constellation] puts the concepts, centered about a thing" (*ND*, 162). Hence the constellation moves beyond the declarative as each concept is indebted to the meaning it receives from, and gives to, the others. Sometimes this occurs through illuminations that result from the friction between concepts; or, some thing or possibility might be pressed out of concealment and brought forward through an accumulation of overlapping thoughts. The latter event is exemplified by Adorno and Horkheimer's treatment of the culture industry; the former by the conflicting insights that they bring to bear on phenomena like *mana*, God, grace, or the self and reflection in the *Odyssey*. Throughout *Negative Dialectics*, Adorno transfigures concepts by juxtaposing them, illuminating the strengths and weaknesses of each through relations of unwonted overlap and tension. Thus he transfigures concepts like identity and nonidentity, appearance and essence, natural and historical, totality and infinity, seriousness and clowning, static and dynamic, meditation and immediacy, equality and inequality, universal and particular, thought and rhetoric, immanence and exteriority. The meaning of each term comes to be in relation to what it is not. For example, "when a category changes [as in negative dialectics] a change occurs in the constellation of all categories, and thus again in each one. Paradigmatically . . . [essence and appearance] come from philosophical tradition and are maintained in negative dialectics, *but their directional tendency is reversed*" (166–67). Furthermore, each of these binary juxtapositions is brought into a subtle relation with the others in order to compose a denser experience of the world—its closures, possibilities, and dangers.

Second, the constellation *manifests* what Adorno (following Max Weber) calls "composing," and through it he seeks to *exemplify* the texture and activity of thinking: gathering concepts around an elusive object; piling up overlapping concepts to evoke a salient aspect of things; juxtaposing concepts in tension with one another to call them into question and illuminate paradoxes or weaknesses concealed when single concepts are thought in isolation; placing concepts in strange relations which change the "directional tendencies" of those concepts' meaning. It is this movement that *manifests* what Adorno calls the "morality of thinking."[39]

Third, constellational writing aims to *solicit* the activity it exemplifies; and this solicitation itself cannot remain simply declarative, given the elements of blindness that accompany static declarations. Hence, the constellation must solicit the morality of thinking through the deployment of vibrant tensions, agonizing and shocking paradoxes among its diverse points. These tensions tend to pull the reader in diverse directions of concern which are conducive to receptive generosity and which simultaneously unsettle thought so that it must begin actively *thinking*.

To help evoke this idea of soliciting constellations, one can extend the image of constellations as prisms that catch the multiplicitous colors of the world. One can imagine prisms casting seductive colors around *selves*—colors that illuminate and intensify, that render distinct and compelling a set of discrepant concerns for people situated within these constellations or under their sway—in order to enliven the morality of thinking. By drawing our concern and desire in diverse and important directions, by assembling around selves a constellation of seductive colors, each of which is too vivid and compelling to ignore or wholly resist, yet few of which fit easily or sleepily with the others, Adorno keeps our senses and thought moving, straining, refocusing to engender a generous comportment in the self's relation to others and otherness. Adorno tries not to muddle and immobilize the ethical dimension (though that is a risk he runs) but rather to animate ethical activity in light of complexities that condition the possibility of receptive generosity. Thus, Adorno offers an ethics of constellational solicitations in contrast with the categorical imperative of the sovereign subject or a code of directions to be unquestioningly followed.

This triple sense of "constellation"—active interpretation, exemplification, and solicitation—adds a complexity to Adorno's work. The interpretive and solicitive dimensions are intertwined in relations of mutual reinforcement in the sense that the interpretation of selves and the world emerges from a thought solicited in particular ways, while this thought in turn is solicited most powerfully from a particular constellation of interpretations.

In this context, let us discuss Adorno's juxtaposition of utopic reconciliation and ineliminable violence. This juxtaposition allows us to exemplify one agonistic combination of thoughts in the constellation he composes in response to the problem of soliciting a morality of thinking in the midst of our tragic tendencies. In addition, it helps make sense of the voice of absolute peace and reconciliation in a philosophy that embraces a dynamic *agon*. Throughout the following discussion, keep in mind Adorno's claim that "attitudes of consciousness are linked by criticizing one another, not by compromising" (*ND*, 131).

At one pole of his thinking, Adorno employs the idea of reconciliation to gesture toward radical transcendence. Gathering and condensing some of Adorno's reflections, we find the following sorts of striking pronouncements. Reconcilement is portrayed as an utterly nonencroaching state of peaceful coexistence and mutual affirmation of difference. Beyond the agonistic generosity of negative dialectics, "reconcilement would release the nonidentical, would rid it of coercion, including spiritualized coercions; it would open the road to the multiplicity of different things . . . the thought of the many as no longer inimical" (*ND*, 6). "Utopia would be above identity and above contradiction . . . a togetherness of diversity" (150). In a situation beyond all relinquishment, "the subject's nonidentity without sacrifice would be utopian" (28). It would be everywhere "the unconfined," symbolized by "a particular free[ing] itself without in turn, by its own particularity, confining others" (306). In short, we would have a "future mankind, pacified and free" (*AT*, 278).

Much that is tacit in these passages is articulated more explicitly elsewhere. For example: "Anti-morality, in rejecting what is immoral in morality, repression, inherits morality's deepest concern: that with all limitations all violence too should be abolished" (*MM*, 95). This concern of morality is for a situation in which morality itself is transcended, one that "would no longer require either repression or morality" (*ND*, 285). Evoking a "freedom . . . without impairment [which] can only be achieved under conditions of unfettered plenty" (218), Adorno pushes the idea of a world without scarcity into the hyper-surreal, with the thought of a "utopia in blind somatic pleasure, which, satisfying the ultimate intention, is intentionless" (*MM*, 61). He imagines a world of post-scarcity, which not only abolishes every trace of "extant suffering but revokes the suffering that is irrevocably past" (*ND*, 403). Opening the floodgates of possibility, Adorno, with modern art, "holds fast to the idea of reconciliation . . . the belief that this earth here, now and immediately could, in virtue of the present potential of the forces of production, become a paradise" (*AT*, 48). In "a solidarity that is transparent to itself and *all the living*" (*ND*, 204), where "duty has the lightness of holiday play" (*MM*, 112), where "eternal peace . . . out of freedom, leaves possibilities unused," we might find ourselves "*rien faire comme une bête*, lying on water and looking peacefully at the sky, 'being, nothing else, without further definition and fulfillment'" (156–57). And in the face of the question whether this utopia does not run counter to life itself, Adorno himself draws the conclusion in no uncertain terms: "To hate destructiveness, one must hate life as well: only death is

an image of undistorted life. . . . [N]ihilistic revulsion . . . is not merely
the psychological, but the objective condition of humanism as Utopia"
(78).

Cramming together these dispersed passages suggests an Adorno
who yearns for total revolution and absolute freedom. Yet these passages do not appear in Adorno's texts in uninterrupted proximity with
one another. My condensation brings into focus *one point* in an agonistic constellation: an element which is articulated in, and draws its
meaning from, its relations with other elements in the constellation.
This soliciting constellation transfigures these longings so that their
"directional tendency" aims not at "total revolution" (nor total despair
at the failure to attain it) but at something quite different. Before passing to ideas that are agonistically juxtaposed with those of utopia, however, let us dwell further on the utopic moment.

With the idea of total reconciliation, Adorno establishes a star in his
ethical constellation that shines from an infinite (and infinitely critical)
distance. In this distant light our identity with the extant world is
loosened, and we are drawn by a remote shimmering—a far-off vision—that opens critically disturbing reflections on the coercive dimensions of our world. This star is not intended to render parts of the
world problematic in a balanced or moderate manner. Rather, it engenders an agonizing discontent that works its way through remediable
social-historical ills toward the fundamental finitude of human life itself: we are limited in our capacities; our being and our choices tragically preclude other possibilities; "life purely as a fact will strangle
other life" (*ND*, 364); we cannot will backward and revoke tragedies.

Adorno took to heart Hegel's thoughts on how a conception of absolute freedom leads to an absolute terror that militantly rages against all
incarnations as heteronomy. And yet, Adorno does not exclude this
voice (radicalized further to affirm uncoerced diversity) in favor of a
more "realistic" and more complacent articulation of freedom. For,
Adorno claims, this radical voice provides one way to listen to the subjugative tragic dimensions of our life. Freedom and well-being depend
on the intransigent distance that this voice provides as a counterforce
against somnambulistic assimilation. This dimension of Adorno's thinking does not allow complacency the slightest foothold. We benefit from
this utopian voice because the finitude, encroachment, and subjugation
that are a fundamental part of life—let alone social class, patriarchy,
the domination of nature, the culture industry—tend toward a conceal-

ment which proliferates unless we think strenuously against the grain of our thought and being, pulled and illuminated partly by the distant light of an utterly transcendent reconciliation. "No light falls on men and things without reflecting transcendence" (*ND*, 404), in which the contingencies of "the possible" might be critically explored in order to open and be opened by "possibility." For all these reasons, Adorno concludes *Minima Moralia* with this sentence: "But beside the demand thus placed on thought, the question of the reality or unreality of redemption itself hardly matters" (247). This line is crucial because it locates the truth of redemption not in some secret eschatological ontology, but in its animating and demanding capacity as part of a constellation of moral solicitations. When the meaning of redemption is thus explicitly shifted from truth beyond time to the living movement of receptive generosity, it "changes its directional tendency": it enlivens thinking, while checking the despair that can accompany the idea of total reconciliation when taken literally and yet seen as impossible. Let us explore this idea further, returning to the broader constellation within which the utopic moment is situated.

The idea of reconciliation is only one point in Adorno's ethical constellation. Or, shifting stellar imagery here to emphasize the juxtapositional quality of his thinking, it is only one *pole* of a star within this constellation which emits light through the tensions between poles. At the other pole we find Adorno repeatedly reminding us of our finitude. "We have no type of cognition at our disposal that differs *absolutely* from the disposing type" (*ND*, 15, my emphasis, to highlight the possibilities of significant variations of degree), the type that engenders damaging identity. Negation and encroachment cling to thought and being, and we can attempt to bestow this structure with a different direction only by means of negation itself. Thus, on the one hand, "no light falls . . . without reflecting transcendence" (404); but, on the other hand, the instant the transcendence of thought becomes positive for itself such that *it* is what matters, then this transcendence engenders the very oblivion to the finitude and specificity of this world that it seeks to escape. The utopian moment keeps soliciting our thoughts toward an utterly sublime exteriority, while the point of tragic finitude in Adorno's constellation never quits beckoning us toward the remembrance that "there is no peeping out. What would lie in the beyond makes its appearance only in the materials and categories within"

(140). And these materials and categories never entirely shed a dimension of blindness and damage.

The point in Adorno's constellation that solicits a recollection of tragic finitude works throughout his aesthetic theory as well as in his reflections on thinking and being. Art is animated by a desire to transcend subjugative relations, both among beings and between particulars and larger social processes: in part, it strives to exemplify reconciled relations. Yet art is also animated by a resolute awareness that "harmony is unattainable, given the strict criteria of what harmony is supposed to be" (*AT*, 161). Adorno repeatedly notes that art "fractures"— no matter how hard it tries not to—the materials on which it works. He calls for artists not to look away from the unavoidable moment of tragic violence in their own activity: "They must not try to erase the fractures left by the process of integration, preserving instead in the aesthetic whole the traces of those elements which resisted integration" (10). He continually reminds us of the ways in which our encroaching being is utterly incapable of inhabiting the indefinite region of reconciliation toward which the utopian point pulls.

In these reflections which draw us alternately toward pondering our damaging finitude and toward the idea of radical reconciliation, one might discern a longing for total transfiguration that leads inexorably to an endless self-flagellating delineation of the textures of pure futility. Yet that would be to misunderstand the relationship between these discrepant thoughts in Adorno's work. These are not two *positive* self-identical ideas, each of which simply bangs its head against the walls of the other. Rather, each idea is *transfigured* and acquires a "different directional tendency" in its agonistic relation to the other idea. To grasp the morality of thinking that Adorno hopes to depict, exemplify, and solicit, one must pay close attention to these transfigurations.

Underscoring the agonistic character of negative dialectics, Adorno writes: "From philosophy we can obtain nothing positive that would be identical with its construction." Each concept must be transfiguratively reinscribed, in a constellation with that to which it is nonidentical, in order to manifest the negativity of each concept with respect to its own idea. Thus, considering the utopic moment, "the idea of reconcilement forbids the positive positing of reconcilement as a concept" (*ND*, 145). Utopia prohibits its own conceptual positivity because it is at odds with the coercive moment of conceptuality, and as soon as it is positively conceptualized it is marked by the subjugative finitude that it seeks to surpass. The idea of reconcilement evokes the transcendence of the transgression and blindness of this world and, hence, cannot be determinately inscribed here without suffering and perpetrating the violence it yearns to overcome. Nor can it rest content within itself as a celebration of purely transcendent indeterminacy, for it would then easily function as a comforting illusion that draws our attention away from the specific subjugative aspects of this world—once again fundamentally undermining itself.

Rather, the idea of reconcilement is driven according to its deepest meaning to articulate itself in critical relations with a world that is not reconciled. We draw closest to this idea when we take up the task of concrete historical critique and judgment with a receptivity and generosity toward nonidentity. As reconcilement becomes this essentially critical activity, it is transfigured and partakes of the transgressive moment it would transcend. Hence, Adorno writes that while the utopian impulse does not want the colors of the world to fade, this impulse only appears "draped in black": tracing in detail the fractures inflicted on that which resisted integration; "a recollection of the possible with a critical edge against the real" (*AT*, 196). In short, "Dissonance is the truth about harmony" (161).

One can discern an analogous transfiguration of the point of tragic finitude. Just as reconciliation is articulated through agonism with the unreconciled, the transgressive dimension of finitude is also articulated and acquires meaning in relation to that which it is not. Thus, while delineating the inherently transgressive dimension of thinking, Adorno argues that "accompanying irreconcilable thoughts is the hope for reconcilement." The resistance of thought to the pressures of immediacy aims, finally, at the dissolution of pressure, not at a reconstitution of pressure in the form of the subject's hegemony. Adorno therefore draws the violent moment beyond self-celebrating positivity, developing antagonism into antagonism against antagonism (which, Adorno claims, it has always been, in part). He traverses the negative, resistant, transgressive moment of thinking to develop the other within its own impulses: identification not simply in the dominating sense, but in the sense of identifying *with* the other; transgression not simply as an eclipse of the object, but as an effort to "heed a potential that waits in the object" (*ND*, 19).

As reconcilement is transfigured and manifests itself with dirty hands in the agonistic work of critique, so too is tragic finitude transfigured such that the recognition of our violence serves a thinking which seeks to reduce damage and subordinate transgression to receptive generosity. Adorno's understanding of the morality of thinking lies in neither of the extreme points when isolated and abstracted from his ethical constellation of solicitations, for the meaning and function of each is fundamentally transfigured through the agonistic relation into which these points are drawn. The points of Adorno's constellations solicit not only us but each other; or, each other through us. Furthermore, these *transfigured points* solicit our thinking in a light that is less extreme than reconciliation and antagonism in their most vibrant expressions. Indeed, much of Adorno's work is written in this more mediated fashion.

Yet if the constellation engenders this transfigured and subtler sense of both the violence and the reconciliation that sometimes shine with such uncompromisingly harsh light in these texts, why doesn't Adorno focus his *entire* strategy around this agonistic middle ground? Why not omit the discussion of extreme points, which appear alternately to yearn for the unattainable and to position us in a pit of despair?

Adorno's response here hinges on his idea of how energy works in texts. Immediately following his claim that "the idea of reconcilement forbids the positing of reconcilement as a concept," he writes: "And yet, in our critique of idealism we do not dismiss any insight once acquired from the concept by its construction, nor any energy once obtained from the method under the concept's guidance" (*ND*, 145). Similarly, Adorno writes that his "dialectics means to break the compulsion to achieve identity, and to break it by means of the energy stored up in that compulsion and congealed in its objectifications" (157). Language is a network of partially congealed and condensed insights and energies, in the sense that concepts participate both in *opening* or *throwing* us into particular perceptions of the world and in *soliciting* our cognitive activity. Hence their *power*. Negative dialectics works with terms, weaving them into constellations to redirect and intensify energies and illuminations toward the critical exploration of possibilities of giving and receiving.

This insight has far-ranging significance for Adorno, and it helps us make sense of his claims that "only exaggeration is true" (*DE*, 118) and that extremes are linked "by criticizing one another, not by compromising" (*ND*, 31). To pitch his argument solely in terms of the critically transfigured and mediated concepts would be to homogenize and devitalize the ethical constellation in order to neatly contain the employed concepts within the dominant tendencies of negative dialectics.

Adorno does not hesitate to transfigure concepts through juxtaposition with others and through formulations that flow more smoothly within the central currents of negative dialectics. The insight and energy of these more moderate reformulations, however, come not only from within the moderate register but from the more extreme formulations that also frequent Adorno's writings. Given the extent to which concealment accompanies thought, the morality of thinking cannot be sufficiently animated by the transfigured perspectives alone, but also requires (mobilizes and is mobilized by) perspectives that exceed the mainstream of its own "better judgment." Thus, Adorno includes these voices in his constellations in order to charge his text with a vibrance of insight and animative energy. This inclusion allows the agonistically transfigured concepts to draw on the energy of extreme critique and extreme hope, both of which have powerful moments of paradoxical truth in our strange lives. Nothing short of this concentration of intense agonistic energies, on Adorno's reading, is likely to solicit a morality of thinking from beings in our somnambulistic society. A more measured and moderate portrayal would likely silence many valuable insights and impulses and would fail "to unite radicalism with restraint" (*AT*, 168).[40]

Thus, the agonistic tensions between reconcilement and tragic finitude help further delineate the contours of Adorno's ethics of giving and receiving. Generosity involves an effort to give, rooted in a concern both for the qualitative specificities of otherness and for the possibilities of transfiguring others and the self toward freedom, well-being, and ethics. Notions of a radical reconciliation of diversity powerfully enliven this dual concern. Simultaneously, a sharp sense of tragic finitude is vital, as it illuminates the transgressive dimensions of our existence and calls us continually to reflect upon the difficulties in our efforts to receive. It also solicits a restraining perspective on the wilder flights of our generous imagination. Under the sway of these discrepant concerns, we become most capable of sustaining a dialogical relationship with the world, wherein lies the greatest possibility for freedom and ethics. It is precisely this dialogue which manifests a "positive notion of enlightenment."

Of course, Adorno might lend himself, at times, to other interpretations. Are there sections where he falls thoroughly under the spell of a mythical notion of pure reconciliation? Do these notions sometimes align themselves dangerously with demands for total revolution that lead to despair? Are there places where the demands for purity may unwittingly lend themselves to disciplinary tendencies in modern societies? Perhaps. More often, however, I think Adorno is closer to the interpretation I develop here. And if his own writing sometimes harbors unnecessary and undesirable risks and complicities, the predominant streams of his thinking suggest readings and further developments of critical theory in which tighter, subtler, and more explicitly articulated juxtapositions increase our distance from these perils and point to other directions. This at any rate is the path I'm pursuing, the spirit of which I'm calling "Adorno."

The Attractions and Repulsions of Receptive Dialogue

We have seen Habermas's claim that Adorno traversed the aporias of modern subjectivity in a manner that exemplified the profound need to move toward a theory of communicative rationality and ethics. Adorno is said to have been either unaware of the possibilities of a communicative turn, or uninterested, or naively dismissive because he yearned for a fantastic and empty total reconciliation. Thus, he does not help us develop discourse ethics.

Yet this picture misses much of what is most interesting in Adorno. There appears to be an ironic failure of communication—a breakdown of receptivity, reading, and listening—in communicative theorists' efforts to communicate about communication in the work of the first-generation Frankfurt School's most important theorist.

Even if it were correct that Adorno had no taste for dialogical theory and ethics (a view I contest in this chapter), it seems that his many comments on the problems with giving center stage to intersubjective communication, the ideal of universal communicability, and the norm of consensuality merit at least a careful rejoinder. And, too, one expects more subtly articulated responses to Adorno's frequent, if paradoxical, rendering of the relationship between subject and object (both in philosophy and in art) in communicative terms. This rendering poses problems for any claim that Adorno remained within a framework of subjectivity limited to instrumental representation and labor. Similarly, one anticipates some discussion of the dialogical insights that are explicitly articulated in Adorno's work. One expects some discussion of the possible significance of the fact that Adorno's first major published work, *Kierkegaard: Construction of the Aesthetic*, concerns a theorist of dialogue, and that the penultimate section in that work contains provocative reflections on communication. But the pickings are slim.

My argument develops as follows. First, I discuss Adorno's critique of communication in order to sketch a position from which to illuminate (in chapter 3 below) some of the ways in which Adorno problematizes the Habermasian perspective. Second, I discuss Adorno's paradoxical employment of dialogical tropes and dialogical interpretations of the relations between subject and object in order to articulate how he conceives of dialogue in ways that both deepen his critique of communication and allow him to begin to move beyond the problems he identifies. Third, I consider specifically *interhuman* dialogue in light of the broader ideas in my second set of observations, focusing on the extralinguistic somatic dimension of all selves. Fourth, I draw these reflections together in an effort to make sense of Adorno's affirmation of the utter importance of dialogical concerns for art and expression more generally. I follow with observations on the interpretive and ethical centrality of the tension between consensus and the more dissentient impulses and "lingering eye" in Adorno's dialogical ethic.

Adorno frequently criticizes positions that make communication (and more specifically communicability) the underlying principle of our art, philosophy, and relations with others more generally. His critique aims not at particular thinkers but rather at a set of underlying assumptions which he finds increasingly ubiquitous, increasingly the spirit of our age (one might read aspects of Habermas's work as specific articulations of that general spirit).

Sometimes Adorno's attacks are provocative, intended to disrupt a complacent acceptance of the dignity and centrality of communication, such as when he writes: "Without exception, what is called communication nowadays is but the noise that drowns out the silence of the spellbound" (*ND*, 348). Here a defender of communication might respond that Adorno is simply criticizing communication between subjects who are (as Adorno writes in the next sentence) constituted in modes of "pseudoactivity and potential idiocy," leaving untouched the core of the imperative to strive for a reflective consensus. Thus Adorno would appear to be criticizing *communication as such* in a sloppy way, over-generalizing his problems with what advocates of communication would call *bad communication*. From the communication theorists' perspective, these problems are attributable to colonized public spheres, poor individuation, unquestioned tradition-bound norms, and so forth; therefore, they are largely curable through a *good* communicative restructuring of society. And when Adorno (in an aphorism entitled "Downwards, ever downwards") rails against those who take communicability to be a regulative ideal, declaring that such an imperative generates a leveling accusation—namely, that "anyone who, in conversation, talks over the head of even one person, is tactless" (*MM*, 183)—proponents of communicability might respond as follows: communicability is a governing ideal demanding everyone's reflective efforts; it is not a positive fact that can be coercively deployed to truncate present efforts at communication.

Still, even though many of Adorno's criticisms of "communicability" are tied to attacks on the ideological institutions, practices, and culture of "late capitalism," many have a depth that exceeds the confines of such obviously "distorted" communication. Indeed, they aim at the heart of the communicative paradigm as such. Enter Adorno's broader critique: "Direct communicability with everyone is not a criterion of truth. We must resist the all but universal compulsion to confuse the communication of knowledge with knowledge itself. . . . Truth is objective, not plausible" (*ND*, 41).[41] The claim here is not, of course, that we can have pure cognitive access to the "objective." Rather, Adorno's point is a *critical* one: it is the claim that "direct communicability," or "communication of knowledge," or (in more contemporary terms) even the ideal that a statement be deemed worthy of rational consensus, is not identical with or exhaustive of the meaning of truth. Communication ought not be transformed into a seamless reality that refers only to itself and that considers the question of truth only in terms of its own intersubjective processes, exchanges, and ideal outcomes. Rather, truth is related to the nonidentical that exceeds communication; that is to say, as in the previous passage, truth is "objective." For Adorno, then, the establishment of rational communication—even the ideal of communicability as the centerpiece of our understanding of truth—is already a foreclosure, an ideological constitution of our idea of truth, a mythical understanding of what we are and what we ought to be doing in our relationship to others and the world.

Adorno's critique, not surprisingly, draws on ways in which communication is entwined with a world that exceeds it, a world that strips the processes and ideals of communication of their self-referential status. First, we always communicate about a world we cannot communicatively subsume. If it is not to lapse into empty tautology, communication must engage the nonidentical within and outside us that resists exhaustive identification within our concepts. Of course, whatever truth we can grasp always becomes present through language and is thus potentially communicable, but our language and communication are not exhaustive of truth. Rather, truth emerges in language (in partial and often paradoxical ways) through language's strange relation to that which is not language. To the extent to which we systematically conceal the relation of language to the extralinguistic by understanding knowledge primarily in terms of universal communicability or consensus, we hamper a "positive notion of enlightenment." Furthermore, the ideal that the world ought to be capturable within a consensus fails to express the openness and dynamic possibility of our relation to nonidentity. Second, communication is entwined with the nonidentical world insofar as the latter always already animates and lies within our communications in ways that are significantly opaque to us. While our bodies are densely mediated by discourse, they are not reducible to discourse either in their specificity or their possibility. This living "physicality" is at "the core of cognition," an otherness that no subject and no intersubjective communication can ever entirely make its own (*ND*, 193). If truth is comprehended primarily in terms of communicability or the ideal of consensus, we conceal an awareness of the way nonidentity always already infuses our intersubjectivity behind our backs. We thus preclude much of the critical reflection fundamental to agonistic generosity. When the ideals of communicability and consensuality are made the autonomous foundation of our knowledge, without a sense of the paradoxical resistance of nonidentity, an insistent demand arises—that our utterances always be unambiguous and entirely straightforward; that they claim totally to understand themselves, avoid paradoxicality and unresolved tensions, contain themselves within the realm of a sovereign universal communicability. The insistent demand arises that our identities conceal their relations with the nonidentical and seal their inherently ambiguous and porous boundaries. Yet this constriction of language weakens our impulse and capacities to engage nonidentity in a more open-ended manner.

Intelligence and receptive generosity become more likely when an awareness of the relation between communication and the extralinguistic becomes vital to our dialogues. When we are aware that the world is "lacking precisely the identity surrogated by thought—[that the world] is contradictory and resists any attempt at unanimous interpretation"— then the regulative demands of unanimity can be loosened, juxtaposed with, and contested by thought that is solicited by a "lingering eye" toward the elusive specificities, paradoxes, and possibilities of the nonidentical (*ND*, 144). This type of thinking risks and sometimes even provokes a profound dissent, which accompanies and creates space for the idiosyncrasies of specific perceptions and thoughts, in order to help express aspects of the distinct otherness of the world. If the world and we ourselves are often contradictory, paradoxical, and manifold, then why embrace a singular assumption that all this could be universally communicated? Perhaps aspects of nonidentity might appear significantly different to different people? And what would it mean to have a consensus about something like a paradox or an openness of possibility—qualities which in their very essence resist definitive communication and which always call for more words, sometimes silence?

This more lingering and more dissenting regulative idea is of no less fundamental importance, nor more transitory, than consensuality. It stands in stark contrast to "the liberal fiction of universal communicability"—which parades itself as "intellectual honesty" but restricts thought to whatever is duplicable within a seamless universal logic which denies that "the value of a thought is [always] measured by its distance from the continuity of the familiar." For Adorno, if we are to cultivate the receptivity that is an integral part of enlightenment, consensuality must be contested by a "lingering eye" that risks and provokes dissent; for "knowledge comes to us through a network of prejudices, opinions, innervations, self-corrections, presuppositions and exaggerations, in short through the dense, firmly founded but by no means uniformly transparent medium of experience" (*MM*, 80).

Extending his critique, Adorno writes: "What is called 'communication' today is the adaptation of spirit to useful aims . . ." (*AT*, 109). Besides addressing issues concerning systems' colonization of the public sphere, Adorno claims that consensuality and "communicability" are posited as *the* defining ideals of "communication" when the imperative to coordinate human interaction and solve problems for useful human purposes comes to override all other aims. When the pressures involved in securing human utility and coordination are, or are taken to be, sufficiently strong, the significantly question-precluding demand for consensuality truncates the dialogical relation between language and that which exceeds language. When communicability advances beyond being *a* demand, and becomes the singular and exhaustive *definition* of communication, this closure gains an impenetrability. These pressures and this definition of communication eclipse our awareness of nonidentity, and they are in turn reinforced by a sensibility blind to the questions and resistances of nonidentity. Certainly a distinction remains between narrowly instrumental communication, which concerns only the means that might best facilitate already determined ends, and a broader communication which concerns questions of ends as well.[42] However, Adorno's point seems to be that a communication subordinated to the imperatives stemming from demands that it be "useful" and "do something" in a relatively immediate way severely truncates the less-pressured (or differently pressured) questioning which is integral to a more intelligent articulation of ends. Hence, even when the direct pressures on communication to be useful are slackened somewhat to permit questions concerning ends, insofar as the communication remains structured and governed by a meta-instrumentality (i.e., imperatives rooted in the demands of utility) it will likely perpetuate blindnesses to nonidentity that ought to be viewed as a *devitalization* and *reduction* of communication—or at least viewed as *only one modality* of communication, rather than its singular essence. Under the pressure of these instrumental "constraints" to decide, oblique approaches to that which is elusive come to be censured. It becomes increasingly difficult to linger with the specificities, distances, ambiguities, and paradoxes of the resistant world around us—all of which are crucial for understanding and ethically engaging others and otherness.

Adorno depicts this subjugative-"communicative" situation with the following analogy: "The straight line is now regarded as the shortest distance between two people, as if they were points. Just as nowadays house-walls are in one piece, so the mortar between people is replaced by the pressure holding them together. Anything different is simply no longer understood" (*MM*, 41).[43] The more we think within the "pressures" of this meta-instrumentality, the more likely we are to come to understand *communication itself* as essentially regulated and defined by the *singular* "straight line" imperative toward consensuality. The latter imperative is in turn too weighty to allow tarrying with the negative in order to articulate, however obliquely and incompletely, the often translucent realm of experience. And so the activity of negative dialectics, which might critically illuminate a singularly insistent consensuality and contest its sovereignty with other beckonings is precluded.

Adorno's description of "what is called communication" was quoted above only in part: "the adaptation of spirit to useful aims. . . ." It continues: "and, worse, to commodity fetishism" (*AT*, 109). This phrase suggests the ways in which imperatives and systems of instrumentality tend to engender relations of equivalence and exchange in order to facilitate their operations. On Adorno's reading, equivalence goes back at least as far as the representational equivalences involved in instrumental sacrifice within "primitive" societies. However, as instrumentality is embodied in self-proliferating systems within capitalist societies, exchange and the law of equivalence penetrate the lifeworld more deeply and extensively. The spirit of this equivalence works its way into our fundamental understandings of communication as defined by the singular a priori demand that truth don the garb of essential exchangeability.

Adorno thinks that this sort of exchange almost always has damaging implications. Nor is such exchange usually the basis of a real gift, as he elucidates in the aphorism "Articles may not be exchanged," discussed above. Adorno sometimes expresses this idea with sarcastic exaggeration: "as if the intent to address men, to adjust to them, did not rob them of what is their due even if they believe the contrary" (*ND*, 368). This idea, repeated throughout his work, aims to expand the space in which to linger with the nonidentical and to dissent from the demands for "communicability." Adorno suggests that it might be better to faithfully present an insight, even if it means that one will fall far outside the realm of "the communicable." This impulse to dissent, to linger in ambiguity, is animated by a sense of the blindness involved in a singularly consensual striving—and, ultimately, by an agonistic idea of generosity that seeks to give people "their due" in part through radical critique and the presentation of recalcitrant differences. Thus, Adorno's critique of "what is called communication" hinges on the ways in which a singular consensuality obfuscates and truncates efforts to pursue the often more circuitous dialogue with that which always remains partly nonidentical. To offset this insistence, Adorno cultivates an insistent concern for the extralinguistic.

Adorno expresses this concern as the effort "to counter Wittgenstein by uttering the unutterable," the effort to employ the linguistic in order to move beyond the linguistic (*ND*, 9). Though Adorno sometimes criticizes Wittgenstein, as when he claims that Wittgenstein "fails to see the constitutive relationship between language and the extra-linguistic" (*AT*, 439), he largely agrees with the latter's insight into the limits of language. Adorno concurs that philosophy's effort to utter the unutterable successfully—such a contradictory challenge—is *"doubtful as ever."* Still, it "is one of philosophy's inalienable features and part of the naiveté that ails it. Otherwise it must capitulate, and the human mind with it" (*ND*, 9, my emphasis). Our efforts to participate in a dialogue with the nonidentical are least prone to blindness and damage when we dwell between, and are pulled discrepantly by (Adorno speaks of "a dialectical tension"), the contrasting solicitations of Walter Benjamin's call to "direct words intensely toward the innermost core of silence," on the one hand, and Wittgenstein's "ontological asceticism of language," on the other (*AT*, 292–93).[44]

The dialogical trajectory of Adorno's agonistic generosity finds expression when he writes: "If the thought really yielded to the object . . . the very objects would start talking under the lingering eye" (*ND*, 27–28). He describes constellations as "interventions" through which objects might "come to speak" (29). Yet if one powerful moment of an ethics of receptive generosity pulls us away from the concern for "direct communicability" with other people and turns us toward speaking, writing, singing, or painting in a receptive effort to give speech to the speechless (in ourselves, others, the nonhuman), Adorno does not solicit this move as if to embrace an *easy* truth or practice. Rather, he discovers the presence of an otherness that seems to be even more oblivious to being received and articulated *by him* than he at times is to being received *by others*. The world that solicits Adorno seems often to inhale its own voice into silence.

Adorno develops these paradoxes in *Aesthetic Theory*. While art and philosophy certainly articulate the relationships between identity and nonidentity in markedly different ways, Adorno's comments here have a general significance that transcends his discussion of art.

As we attempt receptively to give voice to a world of extralinguistic otherness, we sense ourselves to be in the midst of that which "is undefinable in the ordinary sense of conceptualizing something because its substance is precisely its non-generalizability, its non-conceptuality."

And yet, this "essential indeterminacy" does not have the quality of a noumenal thing in itself which resists absolutely all efforts to mark and be marked by it. Nor is this indeterminacy purely plastic, capable of being articulated equally well by each of our efforts to express it. Rather, when we experience the indeterminacy of otherness, as in our experience of natural beauty, we experience "the precedence of the object in subjective experience" (*AT*, 104) in a way that is "hostile to all definition" (107), precisely through nonidentity's mode of appearing simultaneously "as authoritatively valid and as incomprehensible" (105). We discover the "silence of its language," the silence of that which is other than nothing (102). Like the beautiful in nature, nonidentity "is like a spark flashing momentarily and disappearing as soon as one tries to get a hold of it" (107), neither an identical presence nor an utter absence. It is always "appearing to say more than [it] is" (116).

Solicited by this strange sense of silent and incomprehensible authority, this precedence of otherness, we become engaged in thinking and art, in various efforts to "wrest this 'plus' from its contingent setting" (*AT*, 116). In order to negotiate these paradoxes, our own activities come mimetically to resemble our experiences of the movements of otherness repeatedly coming forth and retreating.[45] We seek difference through movements that go toward the world and then retreat in rejection of their own crudity. Thus it is that philosophy and art "partake of morality" (329). Moving in their very different ways, philosophy and art agonistically elaborate each other in "a gesture-like grab for reality, only to draw back violently as [they] touch reality." Art's "letters are like hieroglyphs of this dual kind of motion" (399), and "the self-criticism of reason is its truest morality" (*MM*, 126).

To understand oneself as solicited and resisted in this way is to be drawn into an unending dialogical effort that is only partly grasped in terms of striving to achieve a consensus with nonidentity. Yet consensuality clearly does express an important *aspect* of this effort (as suggested earlier, in the discussion of "identifying with") insofar as we are animated in part by a desire to mark, and be marked by, the world in an agreement that recognizes diverse specificities without illusion or hostility.

Hence, affirming a consensual moment, Adorno writes that despite the problems regarding nonidentity which are engendered by the Hegelian notion of "system," nevertheless its accompanying themes of "unity and unanimity are at the same time an oblique projection of

pacified, no longer antagonistic conditions" (*ND*, 24). Rather than rejecting the ideal of consensuality, Adorno cultivates it as one aspect of
our relation with nonidentity. And yet, our essentially agonistic relationship with a world so frequently fraught with difficulties of violence
and arbitrary constraint also engenders an inextinguishable dissenting
point in the ethical constellation guiding our dialogical strivings with
the nonidentical. The efforts of receptive generosity to ameliorate
senseless suffering and to open, cultivate, and care for the rich possibilities of the world require both consensual and dissentient aspirations. Indeed, every impulse to dissent harbors a yearning for a better
consensus, for a state of things more worthy of diverse beings' mutual
recognition of their togetherness and distinction. But this does not establish the sovereignty or primacy of consensuality, for every consensus
does some damage to the multifarious specificities and indeterminate
possibilities of those who are, and that which is, involved. Every order
to some extent stands in the way of new efforts of receptive generosity.
Hence, consensual striving for a situation beyond subjugation harbors a
yearning for a state of things where dissent is solicited and vibrant. For
Adorno, the possibility of this dialogue with the nonidentical is nurtured in the mutual contestation and solicitation between the impulses
and ideals of consensuality and dissent.

Let us focus directly on specifically interhuman dialogue. Since, for
Adorno, there is no realm of intersubjective communication that is not
permeated and exceeded by nonidentity, the tensions that characterize
his notion of dialogue pull in markedly different directions from those
of an intersubjectivity reduced to the impulse toward universal communicability. The "intersubjective" is always emerging from and returning
to that which it is not. To *approach* other humans generously, and even
receptively, in dialogue involves an aspect of *turning away*, a moment of
rejecting the insistent face which manifests each person's immediate
sense of himself or herself, the present, and the possible. To be sure,
negative dialectics demands a radical receptivity toward the other's face
and voice that might directly transgress one's own self-enclosure—
one's own insistent face and the faces one puts on the world. At the
same time, to respect the face of the other requires in part one's seeking to see through, around, above, below, and beyond the other's face
and the face the other puts on things.[46] It is necessary to seek to withdraw from these faces, from this intersubjective presence, even to the
point where one risks perceiving and articulating something that might

not be receivable by the other person. To strive to "give another his or her due" is in part to present, with as much integrity as possible, that which she or he appears to efface. It is to strive to *supplement* the imperative of signifying to other subjectivities with the aspiration of a receptivity and voice primarily concerned with whatever is nonidentical to extant subjectivities.[47] It is to seek the faces effaced by the continuities of an insistent intersubjectivity; to be drawn powerfully by Rilke's haunting insight that there is "no place which doesn't have eyes to see you" (*AT*, 164), no place without a nonidentical face which solicits a generosity beyond one's capacities to satisfy fully. Only insofar as we are significantly pulled toward recalling effaced faces within others, ourselves, and the surrounding world might we participate in a more dialogical intelligence, generosity, and freedom.[48]

Hence Adorno writes of the need for philosophy and especially for art to "move away from any concern for a viewer" (*AT*, 136), to "free [themselves] of all concern for the sensibilities of the recipient" (281), to distance themselves from a blind allegiance to "those communicative means that would make them palatable to a larger public," to strive toward a dialogue with that which is nonidentical with "the all-encompassing system of communication" (344). This idea is vital for thinking and art; but, as usual, one misinterprets Adorno if one reifies and isolates it.[49]

Like all of Adorno's concepts, this idea harbors its other within. Hence, this "move away from any concern" with the recipient involves a sort of concern that seeks to be more than a "concern," as Adorno suggests when he writes that art's disregard for the recipient, its "inhumanity alone . . . bespeaks its faith in mankind" (*AT*, 281). Adorno's faith is that human beings are indeed capable of participating in receptive generosity. Yet the damaging enclosures of extant selves and intersubjectivity bind us in patterns of expectation—patterns of "the possible"—that must be transfigured, not simply in order to cultivate a gift but to participate in transfiguring others' receptivity as well. So it is that giving, cultivated partly by means of a retreat from concern with others' extant receptivity, aims at provocations that might solicit a more radical receptivity on the part of the "recipient." Adorno writes that "works of art . . . are not created with the recipient in mind but *seek to confront the viewer* with artistic objectivity" (374, my emphasis). In the attempt to bracket one's concern for the recipient's immediate response, a distance is sought through which to "seek to confront the viewer" with different perceptions and thoughts that might draw the other beyond the directions and limits of a reified "communicability." Adorno elaborates this "turning away" as a transfigurative "turning toward," arguing that the demand that our utterances be readily intelligible to the recipient falsely "treats reception as though it were constant and overlooks the impact that unintelligible works can have on consciousness . . . where the uncommunicable is communicated and where the hold of reified consciousness is thus broken" (280). Similarly, "the only way to get through to reified minds by art is to shock them into realizing the phoniness of what a pseudo-scientific terminology likes to call communication. By the same token, art maintains its integrity only by refusing to go along with communication" (443). Art aims to "get through to" its recipients with a shock that opens others to one's own vision of the nonidentical and illuminates the mythological dimension of "communication"—even as it communicates. "Faith in mankind," as a *commitment*, thus manifests itself not only in the preparation of the gift, but also in tending to the provocation of receptivity beyond "the receivable." As a *hope*, it manifests itself by crediting the receivers—however unlikely—with the capacity to be provoked to span a distance without which many gifts cannot be received.

Thus, the move away from concern for communication is, in part, an oblique concern *for* communication. Other aspects of the "move away," however, are less congruous with the aims of communication with others, even those which are quite oblique. The effort to articulate an aspect of nonidentity as faithfully as possible might provoke a shock that is in tension with the effort to shock a recipient toward a more radical receptivity. One effort is oblivious to the task of communication; the other seeks to communicate the "uncommunicable." Hence the relation between these different concerns is often difficult.

When Adorno writes that "the manner in which art communicates with the outside world is in fact also a lack of communication" (*AT*, 7), he in part evokes the "fractures" that these two demands leave throughout most profound works of art. Often the noncommunicative distance necessary for the gift harbors aspects of the cold distance and oblivion that the philosopher or artist seeks to escape. Thus one must repeatedly and consciously strive, through difficult negotiations, to develop both the more radical "move away" and the "concern for" the recipient so that they might, at least partially, reinforce and reciprocally empower each other. As with agonism and reconciliation, they become mutually transfigurative of each other. In this sense, "turning away," though a moment with some autonomy, is also articulated in juxtaposition with and supplemented by a "turn toward."

In a passage on "significance," Adorno develops the tensions involved in the entwinement of (the best senses of) communication and noncommunication. "Significance," which is vital to thought and art, is essentially ambiguous because

> it refers without distinction to the organization of the subject matter as such and to its communication to the audience. . . . Significance designates the point of equilibrium between reason and communication. It is both right, in that the objective figure, the realized expression, turns outward from itself and speaks, and wrong, in that the figure is corrupted by counting in the interlocutor. Every artistic and even theoretical work must show itself able to meet the danger of such ambiguity. Significant form, however esoteric, makes concessions to consumption; lack of significance is dilettantism by its immanent criteria. Quality is decided by the depth at which the work incorporates the alternatives within itself, and so masters them. (*MM*, 142)

This passage identifies both the dangers of "counting in the interlocutor" as well as the dangers (and impossibility, insofar as works have any significance) of "counting the interlocutor out." The latter alternative risks (following Kant's reflections on taste and genius) a superficiality that is as likely as art governed by "communication" to "say nothing"— give nothing—to the subject matter or to others. Finally, these two gifts are utterly (if agonistically) entwined. Part of the gift one seeks to give the object (be it a thing, nonhuman life, self, or other) is a widespread and desirable transfiguration of others feelings, perceptions, and thoughts regarding it. Art and philosophy, then, must strive to make their expression proliferate beyond the canvas or page. In turn, the gift one strives to give to others includes transfigured perceptions, thoughts, and feelings concerning aspects of nonidentity (including those concerning giving and receiving itself) that provoke and participate in the other's freedom and well-being. Thus, each work must (a) strive to mediate in depth the dangers of counting out one aspect in order to achieve the other and (b) seek out possibilities for a more reciprocal articulation between concerns for the nonidentical subject matter and concerns for communication.

In a social context that Adorno thought tended toward a singular and homogenizing emphasis on formulating one's work to make it immediately palatable to "the audience," he chose to accentuate other solicitations. Indeed, I think that he sometimes leans too far in these other directions, occasionally with regard to thematic content, and not infrequently with regard to his textual practice. Still, his most compelling and dominant position does not negate the concern for reception but, rather, strives to transfigure that concern by juxtaposing it with other concerns. Emphasizing the need for a communicative dimension in art that is directly animated by concerns and responsibilities with respect to the recipients and to art's critical place in a broad social context, Adorno writes: "If [art] tries to stay strictly within its autonomous confines it becomes equally cooptable, living a harmless life in its appointed niche." Thus, "modernism's refusal to communicate is a necessary but not sufficient condition of ideology-free art. Such art also requires vitality of expression . . . tensed so as to articulate the tacit posture of works of art. Expression reveals works to be lacerations inflicted by society; expression is the social ferment added to their autonomous shape." Picasso's *Guernica* exemplifies the mediation of autonomy and expression that transcends a mere refusal to communicate, crystallizing as a poignant statement of social protest. For "the socially critical dimensions of art works are those that hurt, those that bring to light . . . what is wrong with present social conditions. The public outcry evoked by works like *Guernica* is a response to that" (*AT*, 337).[50] It is clear here that for Adorno an expressive concern with giving something *to the recipients* is integral to the most profound and successful works, works that are "significant" in both senses discussed above.[51]

As if to indicate the possibility of a sublime point where communication and "turning away" might reinforce each other, Adorno notes that "the art-form which has from earliest times laid the highest claims to spirituality, as representation of Ideas, [viz.] drama, depends equally, by its innermost presuppositions, on an audience" (*MM*, 222). In *Negative Dialectics*, Adorno writes of "ideas" in the "philosophical tradition" (at its best moments) as "negative signs" that "live in the cavities between what things claim to be and what they are" (150). Ideas seek to express the shimmer of that which transcends the present, and their illumination is negatively developed by critiquing whatever would block the light, the color, of these transcending possibilities. Reading the passage on drama in light of Adorno's interpretation of ideas suggests that if

ideas are successfully to resist both oblivion and assimilation, they require an audience. This is so not simply in an external sense, but by its
"innermost presuppositions," insofar as dramatists recognize that "the
highest" is ultimately inextricable from a capacity to be powerfully received and to proliferate in the emotive, perceptual, cognitive, and
practical life of the others before whom drama is presented. As agonistically distinct as "truth content" and "reception" are, the sublime
point of their union guides the efforts of the artist and theorist to go
beyond their conflictual structure toward a mediation in which they
articulate each other—precisely through a recognition of their moment
of irreducible agonism.

The central, if often oblique, nature of a dialogical orientation in
Adorno's thinking is evident not only in his discussion of artistic and
theoretical "production," but in his reflections on dialogue from the
other side as well, from the perspective of the audience that is to receive these works. If, as we've seen, the value of distance is repeatedly
heralded in Adorno's thoughts concerning artistic and philosophical
work, Adorno just as persistently (as should be clear from my reading
of *Negative Dialectics*) solicits and cultivates our desires and efforts to
receive (as an audience) that which is distant. Against those who grasp
philosophy and art as a possession, against those who demand that philosophy and art remain easily assimilable within the confines of "the
communicable," Adorno emphasizes repeatedly that "estrangement
from the world is a moment of art. If one perceives art as anything
other than strange, one does not perceive it at all" (*AT*, 262–63). Again:
"The subject can become identical with the essence of the art work
only by first confronting it like a stranger, externally, and then trying to
compensate for this strangeness by insinuating its subjectivity into the
work. It should never be forgotten, however, that the objectivity of an
art work is never completely and adequately comprehensible but always
ambiguous" (376). If art and philosophy themselves emerge through
difficult efforts to go beyond oneself, then they in turn—and Adorno
as well—demand similar efforts on the part of their recipients. "Subjectively, art calls for externalization. . . . Art is practical in the sense
that it defines the person who experiences it as *zoon politikón* by forcing
him to step outside of himself" (345).[52]

With this reference, Adorno's reflections open onto the horizon of
numerous political questions that he never pursued in his writing. Before we explore some of these possibilities, however, I shall sketch the

central positions of Jürgen Habermas, who understands his project as
the initiation and articulation (with K. O. Apel) of the Frankfurt
School's "communicative turn." Elaborating this position is helpful, for
it will allow us to compare Adorno's movement toward post-secular
caritas with the leading contemporary paradigm of communicative action and ethics. This encounter will allow me to deepen and clarify a
number of the themes I have begun to develop in Adorno's work, and it
will illuminate a series of difficulties in Habermas's work. Finally,
thinking with and against Habermas, I shall draw some possibilities for
beginning to consider political theory and practice in ways that Adorno
seems to have suggested but never articulated or embraced. Prior to
pursuing any of these paths, though, I should formulate a brief response to the questions raised by Charles Taylor concerning the possibility and strength of nontheological moral sources.

Caritas without God

Taylor's challenge is as powerful as it is clear. "High standards need
strong sources."[53] He wonders whether prominent streams of modernity are not living beyond their moral means when they continue to
demand high moral standards while endangering the sources required
to sustain them. Living beyond our means is dangerous, Taylor argues:
in the midst of a moral deficit, we tend increasingly to sustain our
standards through destructive strategies that produce guilt and self-
condemnation, and we project evil onto others. Without positive
sources of affirmation, high moral demands are perpetuated through
strategies of violent negation with which we are all too familiar.
Though Taylor masterfully surveys and draws on modernity's diverse
articulations of expressivist sources, he questions whether they can survive the demise of the very religions whose aura seemingly made possible their birth. Taylor suspects that some capacity to be moved by a
belief in God is essential for a sustainable morality, and that modern
efforts to articulate morality otherwise are "parasitic" on the theologies
they ignore or explicitly reject.

How capable of responding to Taylor's challenges and suspicions is
my reading of the morality of negative dialectics? A full response to
this question is beyond the scope of the present book, but I can suggest

paths that Adorno might have pursued in such an engagement. My claim is that Adorno's philosophy cultivates strong moral sources and transfigures ethical themes suggested by various traditions (mythical, theological, enlightenment), yet it does so without any parasitism on the theological convictions it explicitly rejects.

Clearly Adorno avoids the increasingly secular providential narratives initiated by the early moderns discussed in Taylor's *Sources of the Self*. Nowhere in Adorno's work does one find ontological "harmony of interest" narratives, and Taylor himself recognizes "post-Schopenhauerian" aspects of Adorno's thinking. Yet could Adorno's position rest on a romanticism that discerns aesthetically expressible whispers of God, in a world that exceeds our reason? Though Adorno draws inspiration from expressivist-romantic (especially Hegelian and Marxian) writers, his claim is not that we can express currents of the divine, an originary innocent logos, an intuitively evocable harmony, and so forth. Indeed, we continually participate in and discover relations that involve—or certainly do not preclude—senseless transgression. Still, Adorno shares with numerous expressivists a sense that something greatly worthy of our embrace can occur in negative-dialectical thinking and art.

Gathering together the themes formulated in this chapter, I see Adorno as supporting this last sense in the following ways. First, he makes a compelling case that *intelligence* rapidly withers in the absence of a negative-dialectical morality of thinking at the edge between self and otherness. Except in a narrowly instrumental sense, intelligence (a sense of who we might become, what is valuable, what we ought to pursue and how, a sense of our limits) enjoys its greatest possibilities when we engage the world with an open-ended receptive generosity. One might even say that the ability to dwell critically and expressively *at this edge* is not just conducive to intelligence, but is also its vital aim. Where this ability is lacking, language tends toward tautology: we tend to construct and reinforce walls of blindness, behind which the world and we ourselves disappear.

Second, Adorno connects himself—however agonistically—to the tradition of those who claim that "freedom is inseparable from enlightened thought." In other words, *freedom* is inseparable from a dialogical comportment with others and the world, inseparable from an ethic that emphasizes giving and receiving. Adorno's texts resound with accounts of the reciprocal blindnesses and subjugations that are engendered when we turn from this path. The moral power of negative dialectics stems simultaneously from articulations of these dangers and from intimations of more desirable possibilities.

Third, something like *richness* is also at stake here. Through the morality of thinking, we participate in elaborating the manifold richness of being—the colors and specificities obscured by more totalizing journeys. Adorno's claim, as we have seen, is *not* that all specificity and manifoldness is enriching (a claim that would indeed harbor the whispers of a benevolent God). Rather, Adorno's sense is that *much* of the otherness that receives specific expression in negative dialectics is enriching (vibrant, animating, enabling, edifying, revealing, moving, provocative, nourishing, disruptive of our myopia . . .) and provides opportunities for agnostic encounters through which we might explore new possibilities for empowering transformations of perception, thought, and being. This sense is far more experiential than it is theological. Not that it does not involve a certain faith. But such faith is akin to Merleau-Ponty's "perceptual faith," rooted in existence, that the world of depth both solicits and will remain beyond the cancellation of each perception. In a similar way, Adorno harbors a faith that, however obliquely, our greatest possibilities for enrichment of both self and the surrounding world lie at the edge where identity opens onto the manifold of nonidentity. Our greatest chance to avoid that devitalization of life which so easily works its way into our practices comes when we patiently position ourselves at the tension-laden edge between identity and nonidentity, dwelling with a sense of unseen dangers and expecting unexpectable richer possibilities. If much that we discover bears the marks of (and might also proliferate) subjugation, Adorno's sense nevertheless is that our agonistic engagements with such beings and practices mark the oblique paths of (and toward) a "life that lives."

All of this is entwined with a morality of receptive generosity, as that which both makes possible and is the highest aim of intelligence, freedom, and wealth. As I emphasized earlier, it is not that the self and its values are prior to receptive generosity and then simply employ it as a means; rather, it is through the morality of giving and receiving that the self and value themselves emerge.

However, Taylor might still question whether this ethic is not "conditional on a vision of human nature in the fullness of its health and strength." Does this ethic move us "to extend help to the irremediably broken?"[54] It remains to be seen whether on this score a post-secular *caritas* can match the Christianity Taylor admires. (Yet, however awe-inspiring, one would have to critically explore Mother Teresa before adopting her practices as an unproblematic telos of caring for the broken.) Here I wish only to suggest that negative dialectics has significant animating power.

This power flows from multiple points in Adorno's ethical constellation, including his understandings of the pervasiveness of "damaged life," of the limited human ability to discern the conditions of successful giving and receiving (or even what constitutes "successful"), and of the extent to which edges (including those between life and death, health and illness, etc.) are pregnant sites for unwonted giving and receiving. Entwined with these understandings, Adorno's ethic solicits a commitment to generosity that exceeds calculations of success regarding particular circumstances. Negative dialectics illuminates receptive generosity as a fundamental condition for, and aim of, life-affirming human life. In this sense it becomes, as a regulative ideal, an aim we must not abandon, even under dire circumstances, even in the face of the "irremediably broken." (And who among us is *not* irremediably broken in many ways, and finally?) Negative dialectics solicits generosity precisely *from* and *in* our damaged condition; it does not flee it. However poor our odds, however immediately doomed I or others may be, for Adorno, there is *no* hope and *no* future without generosity. In this vein, Adorno writes of writing for an "imaginary witness" or a "dead God." As I read Adorno, this witness is imagined as the vantage point of receptive generosity as a regulative ideal before which one stands. One adopts this vantage point out of an overwhelming sense that where it is lacking "life does not live." Adorno's work resounds with stories that attest to this truth.

Of course, a "dead God" lacks many of the powers of a living one, and I will not make light of the challenges this difference poses for negative dialectics. Yet the living God, as Taylor notes, has persistently been accompanied by "poisoned chalices" of sublime proportions. In this context, as we grope for a path away from our worst possibilities, my suspicion is that negative dialectics has much to offer—perhaps even a reverence for otherness that is more powerful, more corporeal, vaster than those afforded by many monotheistic theologies. Or perhaps the highest possibilities lie in the agonistic dialogue between negative dialectics and those theologies? At any rate, we should resist efforts to marginalize Adorno's work at the outset.

CHAPTER THREE

Habermas: Coordination Pressures and the Truncation of Communication

The theory of communicative rationality and discourse ethics is an "unfinished project" which has developed increasing complexity and undergone subtle shifts. Here I focus primarily on central themes in Habermas's recent writings in a critical effort to illuminate his differences from Adorno. In the first section I describe the key arguments in Habermas's work. In the second section I develop a critical perspective on Habermas's position.

Becoming and Being Communicatively Human

To comprehend discourse ethics, we begin by analyzing Habermas's understanding of everyday "normal" communicative action; for discourse, or "argumentative speech," is but "a special case—in fact, a privileged derivative—of action oriented toward reaching understanding." Only by conceiving of the former in terms of the latter "can we understand the true thrust of discourse ethics" (*MCCA*, 130). Indeed, the "justificatory power" of this moral consciousness rooted in discursive procedures "stems in the last analysis from the fact that argumentation is rooted in communicative action. The sought-after moral point of view that precedes all controversies originates in a fundamental reciprocity that is built into action oriented toward reaching understanding" (163). Thus if we pass over the character, scope, and depth of everyday communicative action, we will largely miss Habermas's own sense of the "true thrust" and "power" of his position.

This interpretive error underpins one aspect of Albrecht Wellmer's critique of discourse ethics. While Wellmer does not deny that "the practice of arguing is . . . imbued with moral obligations," he questions whether norms of argument "actually betoken obligations of a moral nature," since they say nothing about whether we should enter into a discursive relationship with an other to begin with.[1] Yet, for Habermas, the general moral import of discourse ethics hinges on its derivative status vis-à-vis everyday communicative practice, combined with the utter centrality of that practice in his narrative of who and what human beings are. Because in his view humans are essentially defined and constituted by communicative action, entering into communicative relations of reciprocity with others cannot be grasped as a "choice." To refuse such relations is to violate the condition of possibility of our very humanity. Ultimately the discourse ethics which is presupposed in argumentation is derived from and rooted in our mundane communicative interaction, and it correspondingly gains its power as it "draws attention to the inescapability of the general presuppositions that *always already* underlie the communicative practice of everyday life" (*MCCA*, 130).

Habermas's most elaborate discussion of how everyday communicative action is central to our becoming and being human lies in his work on G. H. Meade in volume 2 of *The Theory of Communicative Action*. While an exhaustive discussion of Meade's understanding of the evolution of interaction from gestures to signal language to the propositionally differentiated communication that distinguishes humans is beyond my scope here, an overview of a few central points illuminates Habermas's understanding of the deeply communicative essence of human beings.

Following Meade, Habermas understands linguistic communication to be deeply rooted in a development through which behavior and action-coordination, originally governed by instincts and stimulus–response mechanisms, came to be "permeated by language," "symbolically restructured," and governed through a process of *communicative* action (*TCA* 2:24). While Habermas (by way of Wittgenstein, J. L. Austin, and John Searle) supplements Meade's focus on coordination and socialization with an analysis of acts of reaching understanding, the latter is deeply colored by the former insofar as achieving understanding is interpreted fundamentally in the context of a species that depends on communication for its survival and development. While derivative modes of communication become possible as we abstract from the pressures of this constitutive context, for Habermas this context is fundamental insofar as it is either responsible for the character of derivative modes (e.g., argumentation) or greatly marginalizes the significance of those modes which deviate from its orientation (e.g., poetic speech).

Habermas reads Meade to claim that "the pressure to adapt that participants in complex interactions exert upon one another . . . puts a premium on the speed of reaction. An advantage accrues to participants who learn not only to interpret the gestures of others in light of their own instinctually anchored reactions, but even to understand the meaning of their own gestures in light of the expected responses of others" (*TCA* 2:12). Such understanding facilitates mutual adjustment and coordination. Yet insofar as the "responses" of the other are understood simply in terms of an "observer perspective" (i.e., a behaviorist stimulus–response framework), the most that can emerge for the participants in this situation is that A (having internalized B's response) and B have *similar* meanings—meanings which are at most objective copies—not *identical* meanings. And following the direction of Meade's argument on "the pressure to adapt," one can recognize that a maximum speed of reaction and numerous associated advantages accrue not to those who acquire only inductively derived approximations of the other's response to one's gestures, but rather to those who share ascriptions of "identical meaning" (14). The presupposition of this *sameness* of meaning ensures the highest level of mutual understanding, the smoothest coordination of action, and the promise of rationally resolving breakdowns of understanding and interaction through critical exchanges aimed at reestablishing agreement.

The move from internalizing the expected objective response of the other to a more truly *communicative* relation with the other occurs quite easily on Habermas's reading. For once a being attains the first level of learning, "it cannot avoid making the gesture *in the expectation* that it will have a certain meaning for the second organism." This expectation entails an attitude change in which one approaches the other as an interpreting social being, a being to be approached with a "communicative intent" that offers something to be understood, not simply a stimulus for response. This in turn implies a *reciprocal relation* between the two "which is constitutive for participants ascribing to the same gesture an *identical* meaning rather than merely undertaking interpretations that are objectively in agreement" (*TCA* 2:13–14).

Habermas's analysis of identity draws heavily on Wittgenstein's discussion of the notion of a rule. Key here is the internal connection between the identity or sameness of a rule and its intersubjective validity. Essentially, a "rule" is based on the agreement of at least two people on its meaning in multiple situations. The claim to be following a rule always means that one is "exposed" to the possibility of a critique from another that one is violating the rule—"a critique that is in principle open to consensus" (*TCA* 2:18). Now, the identity of the meaning of symbols, on Habermas's reading, is established by nothing other than the rules for their use, and these rules develop in the to-and-fro of reciprocal criticism which each participant internalizes. In this intersubjective context, each participant comes to learn the rules determining how others should receive his or her utterances; through the shared interpretation of rules, *identical* meaning emerges.

This developmental process through which "communication" originates is driven by the pressures of everyday life: we adapt to one another in order to facilitate the speedy and precise coordination of complex interactions under conditions of finitude and scarcity. This pressure pushes us toward similar understandings and, beyond similarity, toward identical-meaning ascriptions. If one imagines a slackening of these pressures, perhaps alternative paths of dialogue and interaction become conceivable. Yet such a relaxation is not the condition of everyday life, on Habermas's reading, and hence is not definitive of normal pressurized communication. While Habermas abstains from empirical examinations of the evolutionary aspect of Meade's theory, the notion of the "pressure to adapt" inherent in situations of everyday coordination is vital throughout his work. It underpins his understanding of the centrality of presuppositions of "identical ascriptions of meaning" in communication, entailing in turn the idea of "discursively redeemable validity claims" which foster such identity as well as the normative reciprocity, universality, and consensualism which this idea implies.

These cognitive capacities, Habermas argues, are inextricably intertwined with particular kinds of relations between selves and within the self. Most fundamentally, these relations are characterized by a reversibility and shifting between first, second, and third person attitudes, rooted in his and Meade's understanding of the pressure-laden character of everyday action-coordination. In such circumstances, each participant takes the attitude of the other, continually shifting in expectation between first and second person (leaving aside the third-person position for now), uttering and then internalizing imagined and extant responses. In this shifting of position, a thoroughgoing reversibility is either achieved or, more often, is the *telos* of such movement: ego and alter strive toward a to-and-fro between perspectives (e.g., regarding a symbol or a situation) in which nothing is lost and nothing gained. *Pure reversibility* as identically shared meaning is the counterfactual presupposition and telos of all our communication. With an expectation born under pressure, ego *presupposes* reversibility insofar as ego presumes to know how alter *should* receive ego's utterance. This presupposition opens ego to alter's substantive alterity, should alter differ and disagree; yet such disappointment is immediately bound up in an effort to shift and exchange positions that *aims at* and is *governed by* the horizon of pure reversibility. Under the pressure to resolve differences and reach agreement, the approach to the other is framed and constrained in terms of "problem-solving." Insofar as all participants' interrelations occur according to this pressurized framework, the space and time for expressions and recognitions of nonidentity are at least significantly squeezed; tarrying with the negative is a burdensome cost; and thoughts of *more fundamental* asymmetry or of recalcitrant (perhaps even desirable) nonidentity are precluded from view.

As noted, this understanding of pressurized communication defines all normal communication for Habermas. Indeed, it is precisely the characteristics of language rooted in pressurized action-coordination that Habermas, in *The Philosophical Discourse of Modernity*, marshals against critics (such as Adorno and especially Derrida) who offer markedly different interpretations of language.

Drawing on Austin and Searle, Habermas further develops the notion of communication forged in the midst of the imperatives of a species dependent on linguistically coordinated actions. Concisely: "Under the pressure for decisions proper to the communicative practice of everyday life, participants are dependent upon agreements that coordinate their actions" (*PDM*, 198, affirmatively quoting Richard Ohmann). These agreements have the greatest power to coordinate action—indeed, they are the *only* specifically linguistic communicative power—insofar as they are perceived by the participants as wholly rational validity claims that are free of the influences of instrumental and strategic manipulation, distortive interests, ideological rhetoric, asymmetrical opportunities for participating in deliberation, and so on. Insofar as agreements are deemed to be polluted, their coordinating power dissolves the instant those who are advantaged either weaken relative to those who are not or turn their backs. Normal speech-acts optimally facilitate our *"carrying on the world's business*—describing, urging, contracting, etc.—" insofar as they are grounded in a striving to reach agreements concerning the objective and normative worlds that can stand up to the ongoing tests posed by the ever-present idealizing supposition of a consensus sustainable through "open criticism on the basis of validity claims" (199). Through the "constraints" of these context-transcending idealizations, everyday communication develops a legitimacy for the participants in which an "elocutionary binding force" provides "a mechanism for coordinating action" (196). (This means that the idea that an agreement is legitimate binds people together in a way that significantly affects action.) It is this character of communication that allows us to transcend strategic action and the fateful world of power, and thus act according to the "unforced force" of obligations based on mutual understanding.

In short, the "pressure to decide" in everyday communicatively coordinated action engenders "constraints" within which participants must strive toward an idealized consensus that facilitates such action. These constraints manifest themselves as a "concern to give one's contribution an informative shape, to say what is relevant, to be straight-forward and to avoid obscure, ambiguous, and prolix utterances." They structurally determine everyday communication such that learning processes with independent logics develop and allow us increasingly to master the world's difficulties through the criticism of false interpretations (*PDM*, 204). The language of "straight-forward" is key here, for it defines the relatively insistent unidirectional striving of pressurized communication.

However, not all speech takes place in this manner. Habermas argues that the pressures of everyday communication can be slackened in two very different ways: one culminating in poetic language and the other in argumentative speech, or discourse. By exploring Habermas's understanding of these possibilities, we can better grasp his position on the broader spectrum of communicative experience (further illuminating the crucial role that "everyday communicative action" plays in constructing and sustaining his theoretical and normative perspective).

Habermas's most illuminating discussion of the poetic suspension of everyday pressures is to be found in *The Philosophical Discourse of Modernity*, in the "Excursus on Leveling the Genre Distinction between Philosophy and Literature," where he argues against what he understands to be the postmodernist attempt (exemplified by Derrida and de Man) "to expand the sovereignty of rhetoric over the realm of the logical in order to resolve the problem [that ensnared Adorno] of confronting the totalizing critique of reason" (188). Because Adorno clung to reason even as he attempted to critique it totally, he found himself perpetually in a self-defeating performative contradiction. On Habermas's reading, Derrida seeks to deconstruct logocentric interpretations of language and dissolve all types of language use and genre distinctions (such as those between philosophy, literature, and literary criticism) into an all-embracing, rhetorically governed context that operates wildly behind our backs, thereby establishing the universal supremacy of the world-disclosive poetic function of language. By so doing, Derrida seeks to undermine reason from an entirely external perspective, thereby avoiding the problem of "performative contradictions."

In opposition to this "counter-enlightenment" perception of language as always under the disruptive sway of poetic *différance*, Habermas aims to show that language use governed by the poetic function is (a) confined to a parasitical, peculiar, and limited domain, which is by no means emblematic of language use in normal everyday situations, and is (b) irrelevant to broad reflections on communication and critiques of reason. Derrida's effort to conflate serious and fictional language use and establish the supremacy of the latter (arguing that both are based on playing roles which are infinitely repeatable under unlimited circumstances and are thus permeated by, and bearers of, fictionality) "obviously presupposes what he wants to prove; that the conventions upon which communication is based are not only symbolic but fictional" (*PDM*, 195). Yet the distinction between serious versus fictional discourse does not dissolve into the ubiquitous fact of repeatability; rather, as Austin argues parallel to Habermas's discussion of Meade, it rests on the difference between the presence or absence of norms of action. The norms that govern normal serious language are grounded in the need to coordinate action that continually arises in everyday practices and that demands our efforts to achieve identical understandings. In theatrical, poetic, and humorous types of discourse this "elocutionary force" (and with it the need to seek a binding consensus) is suspended, and language is removed from its normal functional context.

In normal communication, Habermas claims, the rhetorical, world-disclosive, poetic elements of language permeate throughout. Yet since they are constrained by the consensual pressures of everyday action-coordination, they "play a subordinate and supplementary role" (*PDM*, 200). While the normal everyday employment of the poetic function draws our attention to language itself and illuminates our world, its role is closer to that of servant than it is to the sovereign depicted by Derrida. In brief, "the normal language of everyday life is ineradicably rhetorical; but within the matrix of different linguistic functions (particularly those involved in reaching understanding), the rhetorical elements recede here. The world-disclosive linguistic framework is almost at a standstill in the routines of everyday practice" (209).

As everyday pressures are bracketed and language escapes its normal constraints and coordination functions, the poetic function can "gain primacy and structuring force"; it can "take on a life of its own" in which linguistic reflexivity and artful innovation playfully disclose new worlds in a celebration of wild creativity (*PDM*, 204). Yet this poetically structured language use is *derivative*, in Habermas's view, insofar as normal speech is not and cannot be thus determined: the suspensions that make possible the predominance of the poetic function must themselves shortly be rescinded as the imperatives of action and intersubjectivity reassert their demands at the edge of our very limited poetic spaces. Crucial, for Habermas, is that the poetic be marginalized; for on his view it is a realm where consensuality, rational "discursively redeemable validity claims," and normative principles of reciprocity and universality do not predominate. Discourse ethics would not last long, he fears, in the midst of the convulsions and shudders of poetic laughter and rage. Fortunately, not only is the sovereignty of the poetic not—contra Derrida—"normal," but the realm of normal everyday speech is (or can be) relatively sealed from fundamentally disruptive effects of the rhetorical elements. Thus, Habermas is able to focus on reconstructing the rational consensual character of "the normal," undisturbed by the French and their fellow travelers.

We have seen that the pressure to decide questions in everyday communication (concerning truth, normative rightness, and truthfulness), in order to coordinate actions, brings the autonomy of the poetic to the point where it is "almost at a standstill": The extraworldly, multi-worldly, ambiguous-worldly, wild-worldly, condensed-worldly, un-worldly, and so forth are subordinated to the idealized constraints of consensual strivings that underpin the "testing processes of intramundane practice." If for Habermas's Derrida, "linguistically mediated processes within the world are embedded in a *world-constituting* context that prejudices everything" (*PDM*, 205), for Habermas, the world-constituting dimension of language is normally embedded in an intramundane context of problem-solving and carrying on business that engenders idealizations which "prejudice everything" in the direction of consensuality according to criticizable validity claims. This orientation of "everyday communicative practice makes learning processes possible . . . in relation to which the world-disclosive force of interpreting language has in turn to prove its worth. These learning processes unfold an independent logic that transcends all local constraints" (205).

Of course, Habermas knows that these learning processes and independent logics are perpetually distorted and truncated by a plethora of forces, including asymmetries of power and rhetoric, not to mention the overbearing and truncating aspects of the pressure to decide. If these logics and learning processes are to gain their fullest sway and clarity, it is necessary that the pressures of everyday communication be lowered in a way that does not release us from normal pressure-engendered idealizations but, rather, intensifies them. This occurs when the everyday problem-solving orientation is at once sustained and differentiated ("specialized for experiences and modes of knowledge that can be shaped and worked out within the compass of *one* linguistic function and *one* dimension of validity at a time" [*PDM*, 207]) while the temporal horizon is extended in potential to infinity. This idealized extension makes possible the ideal of an infinitely fine filter for purifying consensus. Thus, pressure-constituted everyday communication is realized in the very specific mode in which everyday pressures are suspended. In such discourse, rhetoric is entirely mastered. Discourses, "too, live off of the illuminating power of metaphorical tropes; but the rhetorical elements, which are by no means expunged, are tamed, as it were, and enlisted for special purposes of problem-solving" (209).[2]

In short, the "full power" and "true thrust" of discourse stems at once from the consensual idealizations that emerge from the pressures of everyday communicative coordination (pressures finally linked to temporal finitude) and, at the same time, from a release from these pressures in such a way that the orientations and idealizations of everyday life can expand toward their maximal realization. Discourse tames the pressures of the context in which its idealizations are born, while simultaneously taming (and then enlisting) the potential wildness that results from a too thorough renunciation of the pressures of action-coordination (including their corresponding idealizations). Discourse both tames the pressures and tames depressurization.

Because consensual rationality is given fullest sway in discourse, it is here that the ethical aspects of our idealizing suppositions are most transparent and easily reflected on. However, as noted, the depth and breadth of discourse ethics stems from the fact that "argumentation is a reflective form of communicative action . . . [which] . . . always already presupposes . . . reciprocity and mutual recognition around which *all* moral ideals revolve in everyday life no less than in philosophical ethics" (*MCCA*, 130). In Habermas's view, the consensual idealizing suppositions entail the position articulated in his essays on discourse ethics. The core of this position consists of two principles. The "principle of discourse ethics" (D) states that "only those norms are valid that meet (or could meet) with the approval of all affected in their capacity as *participants in a practical discourse*." This principle in turn presupposes that it is possible to justify a norm, and this possibility rests on the "principle of universalization" (U), a rule of argumentation requiring that "all affected can accept the consequences and the side effects its *general* observance can be anticipated to have for the satisfaction of *everyone's* interest (and these consequences are preferred to those of known alternative possibilities for regulation)." These principles simply make explicit the normative implications of the consensual striving that is necessary for communicatively coordinated action. Thus, (U) delineates "that standpoint from which one can generalize precisely those norms that can count on universal assent because they perceptibly embody an interest common to all affected." In effect, (U) "constrains *all* affected to adopt the perspectives of *all others* in the balancing of interests" that holds out the highest hope for reaching consensus ("DE," 64–65).

Because these principles articulate the normative idealizations deeply rooted in communicative action, we find it impossible to deny them without committing "performative contradictions." Indeed, Habermas repeatedly deploys the charge of "performative contradiction" against his skeptical critics. By arguing for (D) and (U) by means of this charge, Habermas follows the path of K. O. Apel, Jaakko Hintikka and, ultimately, Descartes. Just as "I think, therefore I am" rests on avoiding the performative contradiction implied between the contrary thesis ("I do not exist") and the necessary "existential assumption" ("I exist") that accompanies any thought whatsoever, similarly, to argue against consensuality (concerning truth or rightness) is to commit a performative contradiction insofar as the act of arguing always already presupposes (in trying to convince someone) the thesis of consensuality ("DE," 80).

Even if one follows this refutation of the skeptical rejection of consensual presuppositions of argumentation, and even if one (with Habermas) treats this *refutation* as an important (though not ultimate) *justification* of the thesis of the unavoidability of consensual presuppositions and (U), serious questions remain concerning the import of this unavoidability, especially for normative claims. If argumentation and its presuppositions are merely marginal or occasional phenomena, then the most one might conclude here is that one ought to guide oneself according to (U) *when arguing*, leaving unaddressed crucial questions about when and whether one ought to choose to engage in argumentation and moral relations to begin with. Thus, (U) might be irrelevant to most human relations.

However, Habermas responds by noting that "the significance of these arguments is proportional to the degree of generality of the discourses that entail substantive normative presuppositions" ("DE," 83). As we have seen, moreover, the presuppositions of discourse (which merely bring into clearest reflective form the idealizations of everyday communication) run deep and broad indeed. Hence Habermas's final maneuver against the skeptic who claims that discourse ethics is at best a very limited domain, and then refuses to enter into argumentation at all, is to reassert the inescapability of discursive moral presuppositions at the most general everyday level: "The skeptic may reject morality, but he cannot reject the ethical substance (*Sittlichkeit*) of the life circumstances in which he spends his working hours. . . . [H]e cannot extricate himself from the communicative practice of everyday life in which he is continually forced to take a position by responding yes or no . . . to the presuppositions of which he remains bound" (100–101).

Thinking that he has made a strong case for (U) and (D), Habermas nevertheless seeks to limit the nature and significance of his claims in some important ways. Distancing himself from Apel, Habermas makes no claim to have provided us with an "ultimate justification."[3] Rather, he attributes to his "weak form of transcendental analysis," which claims merely that there are no alternatives to (U) and (D), only a "quasi-empirical validity" ("DE," 96). Moreover, the explicit reconstruction of our pre-theoretical idealizations (U) and (D) (in contrast to their pre-theoretical operation) remains fallible. Since we have no purely transparent access to these intuitions, they must themselves be subject to arguments of interpretation. Important substantiation of these interpretations can be drawn from the corroborations found in more empirical work done by the likes of G. H. Meade and Lawrence Kohlberg. Finally, Habermas cautions against the thought that we can automatically convert the moral presuppositions of argumentation into rules for regulating all action. While these presuppositions may have great significance for substantive moral and legal reflections, the latter realms require additional arguments and justifications and involve a greater degree of empirical and historical contingency.

Habermas argues that the principles of discourse ethics and universalization are meant to guide participants in argumentation, "with the aim of restoring a consensus that has been disrupted" ("DE," 65–67), an idea to which Habermas repeatedly returns. By elaborating this latter notion, we can gain a clearer sense of the status that the discordance within and between selves seems to acquire in the context of consensual universalistic presuppositions. With the early Hegel, Habermas objects to Kant's understanding that such discordance is attributable to the absence of any preestablished harmony between particular perspectives, bodily impulses, desires, and pleasures, on the one hand, and moral universality on the other. Rejecting the idea of any fundamental aspect of discordance, Habermas argues instead that this alienation stems from the division of an ethical totality. This fateful opposition between particulars, and between the particulars and the universal, "results . . . from the disruption of the conditions of symmetry and of the reciprocal dependencies of an intersubjectively constituted life-context, where one part isolates itself and hence also alienates all other parts from itself and their common life. This act of tearing loose from an intersubjectively shared lifeworld is what first generates the subject–object relationship. It is introduced as an alien element, or at least subsequently, into relationships that by nature follow the structure of mutual understanding among subjects" (*PDM*, 29). Now, if this act of tearing loose is the source of all discordance, then the problem has nothing to do with an *excess* of rationality (over against a nonrational corporeality) but simply with a *deficit*. Contra Adorno and others, reason is perceived here as devoid of any essentially tragic, oblivious, violent moment vis-á-vis particularity. Hence, when the repression of "unconstrained communication and the reciprocal gratification of needs" gives rise to a "causality of fate" operating through "split-off symbols and reified grammatical relations," the path beyond this alienation emerges only when hardened opposites resume their commitment to, and efforts toward, mutual understanding based on a rational consensus (*KHI*, 56–59). Viewed in the light of communicative suppositions, agonism is simply a privative, "fallen" condition to be rehabilitated by rational consensual striving.

Habermas's use of certain phrases when he discusses undistorted intersubjective life (e.g., "relationships that by nature follow the structure of mutual understanding" or "unconstrained communication and the reciprocal gratification of needs") can lead one to one an oversimplified understanding of his position that is easy to dismiss. Habermas might appear to embrace a "natural harmony" thesis that overlooks or dismisses—with an astonishing naïveté—those aspects of unruliness which seem to be so deeply a part of the human condition. Habermas himself might occasionally drift toward such weak articulations of his position. For the most part, however, he avoids ontological harmony claims; rather, as we shall see, he rests his case on the claim that there is nothing fundamental about recalcitrance. For Habermas, recalcitrance is the consequence of a disruptive alienation which can be overcome. Hence Habermas actually embraces "plasticity" rather than "harmony." Even his plasticity claim is often limited to the possibility of communication itself, and Habermas draws back somewhat in the face of deeply sedimented and resistant cultural factors and material conditions (this move is most pronounced in *JA*).

Thus, Habermas's central and more plausible argument is that because our existence and coexistence are deeply communicative, we are ontologically and normatively characterized not by de facto agreement but by a mutual lived commitment (agreement) to coexist through efforts to agree. Our shared understanding is most fundamentally an understanding of our living together via processes through which we strive toward mutual understanding: "an unconstrained dialogic relation" (*KHI*, 59), not necessarily a consensus so intensive and extensive

that nothing remains to be discussed. Similarly, "reciprocal gratification of needs" indicates not so much a condition of thoroughly *achieved*
reciprocity but, rather, a reciprocal commitment to strive in the direction of such reciprocity. So, "restoring a consensus that has been disrupted" means restoring the primacy of our commitments to a discursive mediation guided by the principle of universalization. Hence
Habermas writes about "intact intersubjectivities . . . marked by solidarity, though they need not necessarily be free of conflict" (*PT*, 145–
46); and, in an interview, he describes his position as one combining
autonomy and dependance in a way that "does not exclude conflict,
rather it implies those human forms through which one can survive
conflicts" (*AS*, 126).

In this context, discord seems tolerable to Habermas, so long as it
does not involve us in challenging communication and the principle of
universalization itself. And, emphasizing further the minimalist character of his position, the principle of universalization does not suppose
that there is always a general interest which harmonizes what is optimal
for each person involved. While Habermas sometimes writes of (U) as
a "bridging principle, which makes consensus possible, ensur[ing] that
only those norms are accepted as valid that express a *general will*"
("DE," 63), he does not imply either that one will always find such
norms or that they always exist. Rather, (U) is better understood as a
principle for critically testing and exposing false claims to generalizable
interests, for helping us work—obliquely through falsification—toward
discerning better generalizable interests where they do exist, and for
testing and guiding compromises where compromise appears to be the
best that is possible. For "when only particular interests are at stake,
conflicts of action cannot be settled, even in ideal cases, through argumentation [about a general interest optimal for all], but only through
bargaining and compromise" (*AS*, 176). (Habermas argues that the domain of compromise increases in modern societies with increasing differentiation and the pluralization of lifeworlds.) In such cases, (D) and
(U) are meant to guide compromises in terms of procedure and fairness
(see *MCCA*, 205–6). Now, however, in the effort to satisfy everyone's
interests, "*everyone*" must first be disaggregated to recognize heterogeneity, and the subsequent aggregation has the explicit character of an
approximate compromise of losses and gains among those involved.

Yet if conflict is not *excluded* and bargaining is viewed as often "quite
sufficient," discord is nevertheless a contingency that arises *despite*

Habermasian ethical insistence. Habermasian imperatives drive persistently in the opposite direction toward consensuality. Discord becomes, finally, always a privative mode of being, something to be "survived" and something we should strive to overcome. When differences appear to raise intractable barriers to agreement, we retreat to a minimalist position committed to communicative resolutions of problems and henceforth strive to agree on a package of gains and losses that will allow our differences to coexist without discord and in the least undesirable manner. When directly pushed on the question of the possible merits of discord from within the liberal framework, Habermas does not budge. For example, when Perry Anderson, with J. S. Mill in mind, asks him whether "there might still be a tension between the telos of universal consensus and the human (and epistemological) value of conflict and diversity," Habermas appears either to miss or to parry the question, interpreting it as a "question about the rights of pluralism" (*AS*, 173–75).

Habermas addresses this question, concerning the *fact* and the *right* of diversity and discord, in a direct and more compelling manner than he addresses the question concerning their value. The query comes from those who fear that real differences would be crushed under the pressure of the imperative consensuality and formalistic universalizing impulses of his communicative rationality.[4] Habermas has always viewed this fear as misguided: "Nothing makes me more nervous than the imputation . . . [that] the theory of communicative action . . . proposes, or at least suggests, a rationalistic utopian society. I do not regard the fully transparent society as an ideal."[5]

One of his strongest defenses of this claim is found in those essays in *Postmetaphysical Thinking* which elaborate the insight that "repulsion towards the One and veneration of difference and the Other obscures the dialectical connection between them"(*PT*, 140). This argument unfolds in the context of the theoretical position Habermas develops in conjunction with Meade's work on speaker and hearer perspectives. Let's turn now to these reflections on how personal identity is linked to the cognitive and normative aspects of Meade's theory of communication; by doing so, we shall deepen our understanding of the connections between difference and consensuality in Habermas's work.

I have noted that Habermas's discussion of Meade's theory of communication emphasizes the advantage that accrued to those who developed the capacity to interact on the basis of identically understood meanings. Such relatively speedy and accurate communication is better able to respond to the pressures that accompany complex interactions, thereby proving its worth as a relatively efficient means of social coordination. As briefly mentioned above, the cognitive achievement of mutual understanding is entwined with the ability to shift between first-, second-, and third-person perspectives. This shifting, and its crucial role in the formation of personal identity, is of particular interest here.

The sharing of identical meanings, Habermas claims, emerges through a process in which one "interprets one's own behavior through the behavioral reaction of another." This identical interpretation is not fully accomplished at the level of bodily gestures, for the other's perceptual perspective on my body is too discrepant from my own perspective to establish more than probable similarities. Rather, identical-meaning *interpretations* require "that the gesture interpreted by the other is a vocal gesture." The voice makes possible an exact instantaneous coincidence of perceptual perspectives which is capable of establishing cognitive identity: "With the vocal gesture, which both organisms perceive simultaneously, the actor affects himself at the same time and in the same way as he affects his opposite number. This coincidence is supposed to make it possible for the one organism to have an effect upon itself in the same way as it does upon the other and thereby to learn to perceive itself exactly as it is perceived from the view of the other, as a social object. It learns to understand its own behavior . . . in the light of the other's interpreting behavior reaction" (*PT*, 176).

One's vocal gesture is transformed from being merely a "segment of behavior into a sign substrate—the stimulus turns into a bearer of meaning" (*PT*, 176). What is important at present is how these shifts establish an identity of persons that is fundamentally intertwined with communicative interaction. What is depicted is "how the emergence of an originary self-relation is connected with the transition to an evolutionarily new stage of communication" (177). Self-consciousness is communicatively engendered and originates as the self relates to itself from the perspective of an other. The self always henceforth discovers itself as an interactive partner that "a second ago" addressed it. The self-reflective self is always a "me"—an alter ego addressed by the "I" given in memory. It is this communicative relation (governed by a consensual telos) within the self that, Habermas thinks, allows its self-relation to avoid both the objectifying "analytic of finitude" described by

Foucault and the circular retreats of the self from itself elaborated by Adorno.[6]

Yet, more than the communicative structure of the self-relating self takes shape in this analysis. While the cognitive self-relation allows one to "monitor and control" oneself for purposes of mutual understanding, by itself it is incapable of replacing the "coordinative accomplishments" that were previously rooted in instinctually governed behavior. The latter function is henceforth taken over by "normatively generalized behavioral expectations" that become internalized. Internalization occurs as "taking the other's perspective is extended into role-taking" in which the other's normative expectations are inhabited and gradually become an interior locus of moral self-control: a "me" which responds as "the generalized [normative] other" (PT, 179).

At a formal level—beyond the concrete norms of particular social structures—Habermas centers this narrative of normative internal monitoring and control on the genesis of "accountability." Accountability can be viewed as a structure of the self that is required by any society, but for Habermas (and Meade) what is crucial here is that accountability is a fundamental requirement of the autonomy presupposed in communication. Hence, Habermas writes: "Among the universal and unavoidable presuppositions of action oriented to reaching an understanding is the presupposition that the speaker qua actor lays claim to recognition both as an autonomous will and as an individuated being" (PT, 191). Autonomy is a presupposition because "the speaker certainly could not count on the acceptance of his speech acts if he did not already *presuppose* that the addressee took him seriously as someone who could orient his action with validity claims. The one must have recognized the other as an accountable actor whenever he expects him to take a position with a 'yes' or 'no' to his speech act offers" (190). In communicative relations we must recognize others, and be recognized by others, as autonomous beings who are capable of being more than the effects of our surroundings, beings capable of guiding ourselves according to claims to rationality, or we would be unable communicatively to appeal to, or be appealed to by, others. Accountability is an integral part of autonomy, for each "expects" the other to respond rationally to one's own claims, and each is expected by the other to recount, support, and elaborate the reasons by which one guides one's utterances and actions. Thus each must *own*, be responsible for, and reflectively attempt to defend one's utterances and actions. This ownership, this personhood, is generated and sustained by the self's continual return to itself through the expecting and questioning lenses adopted through interactions with the others: it occurs by means of a self-relation of monitoring and control through which autonomous accountability and accountable autonomy are achieved.

The ethical form of this autonomy is shaped by the principles of discourse ethics and universalization. Yet also intertwined with the structure of autonomous accountability is the flip side of universalization, namely *individuation*, now understood as that process through which the self distinguishes itself from others, marks itself as irreplaceable, and presents and realizes itself along these lines. Individuation is *implied* in communicative actors' inter-relations, insofar as each must view the other as irreplaceable, lest substitution obliterate communicative action altogether by silencing the other. Irreplaceable individualization is *achieved*, however, as the self *takes itself* and *holds itself* together in continual self-reflective acts of encountering itself within itself through first- and second-person relations. This relation establishes the synthetic unity of the self—a unity and synthesis through which the self comes to be as truly distinguished, self-developing, irreplaceable, and realized.

This point is clear when Habermas writes that, unlike the identity of things, personal identity is more than the specific intersection of an infinite number of general determinations. Rather, a person *acquires* an identity only through the synthetic activity of taking and holding together: "A person satisfies the conditions and criteria of identity according to which he can be numerically distinguished from others only when he is in a position to ascribe to himself the relevant predicates. In this respect, the predicative self-identification of a person accomplished at an elementary level is a presupposition for that person's being identifiable by others as a person in general—that is, generically—and as a specific person—that is, numerically" (*TCA* 2:105).

Because this relation of reflective self-possession and predication is *presupposed* by communicative action, it is thrust on people insofar as they seek to communicate. Writes Habermas: "The structure of linguistic intersubjectivity that lays down the communicative roles . . . *forces* the participants, insofar as they want to come to an understanding with one another, to act under the presupposition of responsibility [or "accountability": *Zurechnungsfähigkeit*]" (*TCA* 2:100, my emphasis). Habermas further articulates this reciprocally disciplinary moment of intersubjectivity which fosters self-relationships of monitoring, control, and accountability: "The self of an ethical self-understanding is dependent upon recognition by addressees because it generates itself as a response to demands of an other in the first place. Because others attribute accountability to me, I gradually make myself into the one who I have become in living together with others" (*PT*, 170).

Habermas conceives of self-ownership not as the characteristic of an isolated possessive individual but as something "generated" by and "dependent" on the demands and recognition of others in a communicative context. Now, this might appear to open the genesis and perpetuation of personal identity to some very difficult problems. So long as the self's identity appears to be, and *is*, simply an internalization of the others, of the "generalized other," or of the social structure, its autonomy and distinctness would seem to be little more than a myth. The self's activity might appear limited to proliferating the normalizing gaze of the collective and to internalizing socially "desirable" characteristics, while being fundamentally passive in relation to the desirability of these norms and characteristics. The communicatively engendered self-relation would thus appear as a mechanism of subjugation.

Yet these problems do not necessarily arise and, in fact, Habermas argues, become less threatening the more that human relations actually follow the structure of linguistic intersubjectivity. As increasing rationalization makes possible the development of highly differentiated systems and social relations that place multiple and conflicting demands on selves, and as lifeworld traditions come under attack from colonizing systems and criticism stemming from a normative rationalization that becomes increasingly abstract and detached from unreflective traditional values, both norms *and identity* become rooted in the *idealizing* structures of discourse. Destabilized by the increasing role conflicts of a more differentiated society and by withering traditions, the ego develops a nonconventional identity: not that of a formal, abstract, and isolated subject, but one that is "stabilized in relationships of reciprocal recognition that are at least *anticipated*" (*PT*, 184).

In a manner analogous to the way that both truth and normative claims are generated with an eye toward their worthiness of establishing an idealized consensus, identity comes to be rooted in relation to a "wider commonwealth of rational beings" (*PT*, 185). Identity here refers not simply to one's identity as shaped by the universalistic requirements of communicative ethics, but moreover to the distinct irreplaceable aspect of one's identity, which is forged in light of its possible recognition in an "ideal communication community." And "corresponding to . . . [this] community is an identity that makes possible self-realization on the basis of autonomous action." Yet it is no longer the self's concrete attributes that fundamentally define this identity, but rather the very process of circular self-relations of accountability and control—a process that involves a shifting among various perspectives through which specifically human identity comes to be. Identity in relation to the anticipated community becomes, fundamentally, the *continuity* and *consistency* which is demanded by the structure of intersubjectivity and which makes personal identity possible. Thus, "this identity proves itself in the ability to lend continuity to one's own life history. . . . [This] identity of the [postconventional] ego can then be established only through the abstract ability to satisfy the requirements of consistency, and thereby the conditions of recognition, in the face of incompatible role expectations. . . . The ego-identity of an adult proves its worth in the ability to build up new identities from shattered or superseded identities, and to integrate them with old identities in such a way that the fabric of one's interactions is organized into the unity of a life history that is both unmistakable and accountable" (*TCA* 2:98).

One's individuated identity is (hypothetically) recognized in this anticipated ideal community as one establishes continuity and consistency— even in the midst of shattering events—and creates a responsible life. In Habermas's view, every communicative interaction presupposes this responsibility; yet, in light of an ideal speech community, this demand and need is unleashed from the ideological attributes of the concrete world and self that irrationally preclude the possibility of self-presentation and self-realization within a framework of mutual empathy and respect. Thus unleashed, the synthesizing powers of the communicative self develop a responsibility and accountability that far transcends a given conversation and that extends ideally to the *entire life* of persons mutually recognized. "To the extent that the adult can take over and be responsible for his own biography he can come back to himself in the narratively preserved traces of his own interactions. Only one who takes over his own life history can see in it the realization of his self. Responsibly to take over one's own biography means to get clear about *who one wants to be*, and from this horizon to view the traces of one's own interactions *as if* they were deposited by the actions of a responsible author, of a subject that acted on the basis of a reflective relation to self" (*TCA* 2:99). The self's accountability must not stop at the edges of an argument: such a self would never be realized, nor achieve recognition in ideal circumstances. Of course this striving is constantly upset in real life, but that is no reason to veer from this singular direction.

Habermas, with Meade, claims that the idealized community plays a *revolutionary* role by allowing the individual to "free himself from the fetters of habitual concrete conditions of life" (*TCA* 2:97). Yet, as I shall illustrate shortly, it also becomes apparent that in this lifting of "irrational blockages" the relatively depressurized mutuality of the ideal communication community follows exactly the contours of communication born under pressure. These pressures are perpetuated, leave their traces everywhere, and are never questioned within the idealizations which are to free us of our fetters.

Nevertheless, in Habermas's view, this development of personal identity acts as a co-originary counterpoint to the consensual idealizations that are always presupposed in communication. As we strive for agreement we must simultaneously render ourselves autonomous and distinct, and we must grant these qualities to our interlocutors as well.

With these suppositions—truth, norms, and individual identities which are developed in light of their worthiness of recognition in an idealized community—we are driven to transcend critically those contingencies of history and power that block our development as autonomous communicative beings. Given Habermas's understanding of the intertwining of consensuality and autonomy, his comments on the "dialectical connection" between unity and difference becomes more intelligible, as does his continual insistence that his theory does not lead to a singular focus on consensual transparency. In the idealizing supposition of a consensus open to criticism, the possibility of diverse voices is not repressed but is rather the very condition of legitimacy: "The intersubjectivity of linguistically attained consensus does not eradicate from the accord the differences in speaker perspectives but rather presupposes them as ineliminable" (*PT*, 48). Under the constraint of the pragmatic idealizations of communicative striving toward truth, the "I" is bound to maintain an openness toward the possible criticisms of each of the "others" in the unlimited communication community. Hence, in the insistent consensual demands, Habermas sees the requirements for an inexhaustible openness to the infinite nonidentical "thou's" with whom one coexists.

The different speaker perspectives infuse intersubjective accord with a "porosity" which is more than a counterfactual ideal, for as we open to the others through idealizing presuppositions of identical ascriptions of meaning and agreement, the "shadow of difference is cast" by "*the fact* that the intentions of speakers diverge again and again from the standard meanings" (*PT*, 47–49, my emphasis). As these idealizations gain greater sway in modernity, they participate in a movement toward differentiation: "The transitory unity that is generated in the porous and refracted intersubjectivity of a linguistically mediated consensus not only supports but furthers and accelerates the pluralization of forms of life and the individualization of lifestyles. More discourse means more contradiction and difference. The more abstract the agreements become, the more diverse the disagreements with which we can *nonviolently* live" (140).

Of course, universality and difference are utterly entwined for Adorno, too. Yet, in the connection Habermas describes (i.e., between universality and difference in the communicative rationality and ethics which are said to accomplish the "profane rescue of the nonidentical") the absence of agony and paradox—even of wounds—stands in sharp contrast with the more agitated aporetical compositions of Adorno. Thus, in order to deepen our discussion of these positions, let's turn to a more direct engagement between the two.

Toward a Negative-Dialectical Critique of Communicative Action and Ethics

To begin to understand the differences between Adorno's negative-dialectical dialogue and Habermas's communication, we must move beyond the discussion of Habermas's analysis of the lines, dynamics, pressures, relations, strivings, and tendencies which define the *interior* space and time of communication. We must focus instead on his treatment of that which lies *outside* intersubjectivity, and we must consider the nature of the edge between this "interior" and "exterior." In this light, we can begin to discern the assumptions that underpin and are sustained by his paradigm, and we can see some broader implications of the notions of plasticity, persistent blindnesses, and insistent unidirectional strivings as they operate within the framework of communicative action. In the context of this larger picture, I will conduct some critical interrogations of the Habermasian project which further clarify and illuminate the relative strength of Adorno's position as developed above. Following Habermas's division of that which lies outside the sphere of intersubjectivity into "outer" and "inner" nature, I begin with a discussion of the former.

As is well known, for nearly three decades Habermas has rejected the first-generation critical theorists' diverse hopes and visions of transcending a merely instrumental relationship with nonhuman nature.[7] Since Habermas's most sustained defense of this position is to be found in his discussion of Marx in *Knowledge and Human Interests*, I will focus primarily on that text. Following Marx, Habermas rejects Hegel's philosophy of absolute knowledge and identity on the basis of an understanding that nature is external to consciousness and is "a substratum on which the mind contingently depends" (*KHI*, 26). In light of this given and insurmountable nonidentity between subjectivity and nature, a question emerges concerning their relation. For Habermas (and his Marx) this relation is "rooted in real labor (*Arbeit*) processes," through which the human species "reproduces its life under natural conditions" (27). These processes of labor, which Habermas (though not Marx) views exclusively as "instrumental action," constitute the "conditions of possible objectivity of the objects of experience" (28). Though labor processes undergo historical transformation, the telos of the technical control of nature is invariant and establishes a deeply rooted quasi-transcendental cognitive-perceptual framework (analogous to Kant's faculties) with which nature as an object of experience must conform. In a sense, then, in Habermas's view, we *give* ourselves the experience of nature, insofar as we know it only through the structuring lens of our own imperatives.

Yet this materialist synthesis follows Kant only so far—not simply because we *receive* these instrumental imperatives through our contingently given bodies, but also because we do have an *experience* of nature's "complete otherness," even if only in our encounter with the contingency facticity of nature's processes. If these processes appear only in experience constituted through human instrumental activity, their independence, Habermas claims, is nevertheless revealed in the form of "laws" of nature which, so far as we can discern, are contingently given and must simply be "obeyed" if we are to submit the outer world to technical control. Hence, the limit to instrumental activity appears in the very experience that this activity makes possible: we cannot make the laws. We do not discover our sovereignty as lawgivers, as with Kant, but rather establish it through an instrumental mastery based on conscious reception of the immutable laws of nature, which tell us nothing about radical otherness except that it exists independent of us.

With this particular acknowledgment of an "other in itself," however, the *problem* of receiving is really dissolved. Nature can be understood only through an instrumental lens. We must recognize the independent otherness of nature, but that recognition is rendered meaningless in the radical impenetrability and irrelevance of the essential absence for us of its "in-itselfness." Hegel's seal of absolute identity is broken and replaced by a limited instrumental identity of experience whose boundaries are sealed with a compulsion every bit as powerful as was Hegel's compulsion to subsume. And in this "binding," as Habermas frequently refers to it, the space for other questions and for receptivity toward the nonhuman is squeezed shut (*KHI*, 32–36). In brief, "there is for *this* domain of reality only one *theoretically fruitful* attitude, namely the objectivating attitude" linked to instrumental relations ("RC," 243–44). Similarly, Habermas does not simply see important differences but, rather, a "yawning gap" between discourse ethics and misguided efforts to extend truly ethical relations to the nonhuman.

If the problem of the otherness of outer nature is dissolved in a formula of irrelevance, then the problem of the otherness of inner nature is, as *Knowledge and Human Interests* shows, dissolved in absolute nothingness. Habermas seeks to interpret and transfigure psychoanalysis through the lens of linguistic communication. To maintain the undisturbed primacy of the linguistic, however, he must come to terms with Freud's "internal foreign territory" of instinctual drives and energies. The latter would appear highly recalcitrant to linguistic consciousness, not simply as the resistant objects of self-reflection but also as that which partially governs and distorts consciousness and communication behind our backs.

Habermas claims that, in human beings, the legacy of instinctual drives attached to bodily functions has been supplanted by "a plastic impulse potential, which, while pre-oriented in libidinal and aggressive directions is otherwise undefined, owing to its uncoupling from inherited motor activity" (*KHI*, 239). This impulse potential takes determinate shape only through a process of linguistic articulation "in the form of *need interpretations*" (*KHI*, 241). In other words, Freud's idea of an inner nature of unconscious, yet somehow determinate, "asymbolic

ideas" carried out on a "nonlinguistic substratum" is highly problematic owing to a false "objectivist turn" that reifies processes really
rooted in the dynamics of lifeworld communication and power.

Freud forgets that "we have only *derived* the concept of impulse [or
drive, instinct] privatively from language deformation and behavioral
pathology" (*KHI*, 285). Rather than understand such impulses as independent and determinate objective drives, we ought to conceive of
them as linguistic need interpretations and motives that—having been
split off from and barred entry into the linguistic realm of communication, owing to the requirements of asymmetries of power—acquire a
private life of their own behind the backs of conscious subjects. These
unacknowledgeable need interpretations and symbols then compel consciousness from outside, distorting its intentions, communication, behavior, and so forth. Central here is that the realm of determinate inner
nature, far from being in any way ontologically nonidentical to the
realm of linguistic intersubjectivity, is simply that which has been "excommunicated" from intersubjectivity by repressive power relations.
Therefore, determinate inner nature, at least in principle, is essentially
retrievable and subject to transparent reflection. The linguistic realm,
then, is not engaging something that is in some sense fundamentally
"other"; it is merely retrieving itself.

The path toward retrieval of an inner nature of excommunicated,
linguistically articulated needs and motives requires "a *compensatory
learning process* [including, for Habermas, social change], *which undoes
processes of splitting off.*" This process aims to restore "the virtual totality
. . . represented by the model of pure communication," such that "all
habitual interactions and interpretations relevant to life conduct are
accessible at all times. This is possible on the basis of internalizing the
apparatus of unrestricted ordinary language of uncompelled and public
communication, so that the transparency of recollected life is preserved" (*KHI*, 232–33).

When pushed, Habermas does not seem to view either the development of the self or the processes of self-reflection as capable of actually
achieving transparency. Rather, transparency is simply a transcendent
telos of therapeutic dialogue and self-reflection ("RC," 235). Yet there
are many questions that can be raised concerning the exclusive singularity of this telos of totality and transparency, Habermas's portrayal of
the relative ease of our journey along this path, the lack of *any* accompanying tragic dimension or sense of loss, and so forth. These ques-

tions, of course, are precluded if one accepts as unproblematic Haber-mas's claims concerning the utter absence of a nonlinguistic realm with any determinacy or substantiality that might be other than, and re-sistant to, linguistic intersubjectivity. Given these assumptions, there is in principle, if not a preestablished harmony, at least no basic recal-citrance, discord, or opacity with respect to a reflective linguistic en-gagement with our inner being that aims at transparency and totality. Thus it seems legitimate here to think of inner nature as "communi-catively fluid," as alterable in light of a "discursive formation of will" into which it is drawn as an essentially unproblematic participant since it is always already discourse.[8] If inner nature is *simply* a split-off "inter-pretation," then it would seem particularly suitable to transformation in light of consensual engagements with other interpretations. But what is the status and legitimacy of this narrative?

Though Habermas uses terms such as "quasi-transcendental"[9] and "transcendental-pragmatic" to describe his project, such language tends to conceal, I think, the finitude that he often recognizes as constituting all philosophical reflection. Habermas powerfully articulates the end-less epistemological circle involved in self-reflection, identified in Hegel's critique of Kant's project to establish a transcendental "first philosophy." Habermas also illustrates problems with Hegel's own ef-fort to escape the contingencies of the circle through a philosophy of absolute mind. And when pushed, he acknowledges that his "transcen-dental pragmatic justification" provides no "ultimate justification": it is fallible, and it needs indirect corroboration ("DE," 97–98). Neverthe-less, Habermas clings to the rhetoric of transcendentalism. In part, that is because Habermas carries on the transcendental *aspiration* of discern-ing universal and necessary conditions of possibility for cognition. Yet it is not clear exactly what this means under conditions of infinite cir-cular regress. At present, having recognized an essential fallibility accompanying all theoretical reconstructions of pre-theoretical knowl-edge in argumentation, he seems to have retreated to a vague and problematic claim about a transcendental certainty which is located in the "intuitive knowledge" underpinning communicative action (97). In-tentions aside, however, the discursive effects of the rhetoric of tran-scendentalism are, on the one hand, largely to exaggerate the status, power, and transhistorical qualities of his arguments; and, on the other, to minimize our sense of their finitude, contingency, and uncertainty. This in turn can deflect or dampen the intensity of critical questioning, not only by readers of this rhetoric but by those, such as Habermas, who deploy this rhetoric and who are perhaps carried along and blinded by it as well. Recognizing the dangers of transcendental rhetoric, along with the circularity, fallibility, and finitude of philosophical reflection, perhaps we should temper the language we use to describe our aspirations, so as to bring them into closer accord with the conditions of our pursuit. (Or perhaps we should juxtapose the rhetoric of transcendentalism with rhetoric evoking our finitude.) In this sense, one might describe Habermas's project as a fallible effort to illuminate a portion of the conditions of our cognition from a perspective which is deeply and broadly oriented by a set of personal and historical presuppositions.

Joel Whitebook raises similar questions: "Is Habermas himself not guilty of arbitrarily arresting reflection?" Whitebook suggests that Habermas has marginalized reflections on "the conditions of the possibility of discovering the conditions of possibility of knowledge." He discerns, in Habermas's transcendentalism, more "aesthetic taste" and "judgement" than "emphatic philosophical proof."[10] Yet if this is true and if we therefore bracket the rhetoric of "transcendental" on one side, and that of "skeptic" or "nihilist" on the other, and with a less rhetorical Habermas take "the single most important question" to be which among all contesting positions offers a "better explanation," then we can bring Adorno to bear on Habermas's position in illuminating ways.[11] I begin with some reflections concerning inner and outer nature.

Two things concerning inner nature should be said at the outset. First, we ought to reject Habermas's understanding of the mutually exclusive alternatives for describing inner nature: *either* purely "interactional concepts" *or* concepts that suggest an utterly "physicalistic" or "biological" order of things ("QC," 213). The idea of a realm of split-off symbols forming a kind of inner nature that operates unconsciously with a fatelike causality is certainly not at odds with, or even foreign to, Adorno's theoretical position and specific insights—even if Adorno failed to formulate this idea in as systematic, powerful, and analytically useful a manner as Habermas. Hence, what is at issue here between Adorno and Habermas is a rejection not of Habermas's valuable insights but, rather, of the claim that they exhaustively define inner nature and legitimately preclude the thought of a significant and resistant extralinguistic dimension to selves. It is important to note that Habermas never offers much of an argument for his position. It stands primarily as an assertion. To reject this position, furthermore, it is not necessary to adopt a reified deterministic biologism. Another possibility emerges from Adorno's reflections on the recalcitrant nonidentity of the body that always exceeds but is not totally resistant to our efforts to comprehend and practically modify it.

All of this brings us to a second point. If Habermas offers *any* argument concerning the exclusively linguistic nature of selves, it seems to be that since "we never encounter any needs that are not already interpreted linguistically and symbolically affixed to potential actions" (*KHI*, 285), our concept, and indeed the very *being*, of the extralinguistic aspect to selves must be privative vis-à-vis the linguistic. Adorno agrees with the epistemological part of this claim, as I argued in my discussion of the conceptually immanent and privative character of Adorno's sense of the extralinguistic. Yet as Whitebook suggests, following Freud, there is no reason to suppose a necessary correspondence between "the order of knowing" and "the order of being": the latter remains a distinct question, even when Habermas, Adorno, and Freud agree that we have *cognitive* access only to the linguistically and symbolically mediated.[12]

However, given the extent of our linguistic immanence, how can we make *any* ontological claims about the being or nonbeing of a significant and resistant extralinguistic dimension of selves? In truth, I think we are on the terrain of indirect judgment here. Yet, at the same time, this is an unavoidable question and terrain insofar as we are always thinking and acting on the basis of (at least tacit) thoughts about who we are and about the character and position of linguistic consciousness in this configuration. As we can see in the juxtaposition of Habermas and Freud, our judgment at this basic level greatly influences the ways in which we understand ourselves as well as our sense of what is possible and desirable.[13]

Beyond physicalistic or linguistic reductionism,[14] Adorno draws us to the importance of a "somatic moment" that significantly infuses cognition, is an object of its concern, and yet remains inexhaustibly nonidentical or irreducible to thought. Gathering together his unsystematic reflections into a defense, the following points seem important. First, our very opening onto the world, with which cognition finds itself entwined and to which it is indebted, exceeds linguistic consciousness. To a great extent, that is because our embodiment in the world gives birth to perception in connection with somatic location, energies and bodily habits, impulses, sufferings, sympathies, dispositions, moods, pleasures, passions, and desires. We are always already thrown by historical bodies that exceed linguistic consciousness. Some aspects of the bodily seem quite plastic and transparent, others seem extremely recalcitrant to alteration and even to understanding. And the fact that they are always already there, even in our efforts to understand them, casts a wide shadow of ambiguity and fallibility on the results of these efforts, which would seem to require a greater element of distance from these aspects of the bodily to succeed.

Second, the perceptual field that we face on self-reflection seems to have characteristics of specificity, manifoldness, and openness that are entwined with language but exceed the capacities of linguistic consciousness to exhaust. Like an interpreter before an artwork, the self is itself "subject to a potentially infinite number of postulates, none of which he can satisfy without violating others" (*AT*, 391). No doubt this infinity is significantly owing to the open and manifold dynamic web of language. But, on Adorno's reading, it also seems very much owing, to a discrepancy and nonidentity between linguistic cognition and the bodily-worldly field that language can never completely extinguish. That this field becomes cognizable only with a high degree of determinacy in language does not suggest that it is utterly "plastic" prior to language, or no more than its linguistic "interpretation," for as language brings forth determinate renderings of the world, it endlessly discovers retrospectively aspects of the world which had been concealed and damaged by its concepts. This repeated encounter with language's inadequacy gives rise to a sense of recalcitrant extralinguistic nonidentity that is strong enough to survive the fact that it itself is always born (with determinacy) in a movement *of language* that has partly revealed and thus partly transcended its own limits. Hence, rather than project a totalizing linguistic ontology (which exhibits an exaggerated faith in linguistic consciousness) based on a dubious slide from an "order of knowing" to an "order of being," Adorno projects a sense of an eternally recurring concealment and damage that accompanies language, so that we might better negotiate this condition.

Adorno also resists the notion that outer nature must be singularly reduced to an "in itself" that can be cognitively apprehended only within an instrumental framework. For Adorno, as I have discussed elsewhere[15] and in Chapter 2 above, a thriving freedom, while clearly involving practical and cognitive aspects of control over parts of non-human nature, significantly emerges and is practiced insofar as we fashion our lives in an engagement with our nonhuman surroundings in a manner that aims at increasing our consciousness of the world's otherness. The need for a significant degree of control within our environment does not constitute our *entire cognitive relation* to the nonhuman within a transcendental framework of instrumental mastery. The telos of our lives with respect to the nonhuman is by no means simply given or singular; rather, it is largely a question for us that calls for an ongoing negative-dialectical encounter through which we attempt to discern and articulate where we are, with whom and with what we exist, what we ought to do and what we ought to let be. The nonhuman world provides no clear answers to these questions; but neither is that world simply, as Donna Haraway aptly acknowledges, "a blank page for social inscription," for "the ephemera of discursive production and social construction."[16] "Nature's language is mute," as Adorno writes, "like a spark flashing momentarily and disappearing as soon as one tries to get a hold of it" (*AT*, 107), but it is not *nothing*. It provokes questions, solicits our efforts to articulate responses, and even suggests directions of articulation, though never without ambiguity. And in this never-ending questioning and responding—which itself has a sense in the richness and enlivening indeterminacy of its movement—the senses and purposes of our lives emerge. This is "cognitively unfruitful" only insofar as we can only recognize "fruit" in the relatively closed identity and teleology of a transcendental pragmatic framework. But what if the continual reemergence of questionability is itself fruitful?

Simply to assert, as Habermas does, that the instrumental attitude toward the nonhuman world is alone "cognitively fruitful" is to arbitrarily arrest all but the most technocentric reflections on important questions: how much control, the meaning of control, what kind of control, when not to control, the limits of control, other possibilities, what other beings might provide metaphors through which we can illuminate and shape our lives, and so forth. These questions are vital to our lives, and they involve but exceed questions of instrumentality and interhuman relations. The importance of these questions and our responses to them suggests that we ought to recognize negative-dialectical engagements with outer nature as having a kind of priority over, and a "cognitive fruitfulness" superior to (even if far less definite than), instrumental relations with the nonhuman. For instrumental mastery ought to be carefully framed and limited by other concerns, questions, and understandings.

If our identities are intertwined with nonidentity, as Adorno maintains, then we are called to a negative-dialectical engagement through which our judgment, speech, and action ought repeatedly to be formed and deformed. It is through this practice that we are most capable of living—in the face of the manifold specificity, relations, and paradoxes that both *are* us and *surround* us—in ways which are most likely to reduce the violence to self and otherness that accompanies our narrowing vision, our arbitrarily arrested reflections. Through this practice we both agonistically and cooperatively encounter the vast and potentially enriching diversity of the world and, in this way, might become aware of a different "interest."

If our intelligence, receptive generosity, freedom, ability to resist danger, and thriving hinge on these dialogical engagements at the inexhaustible edge between identity and nonidentity, one can begin to imagine an ethic that exceeds the sovereign bounds of exclusively consensualist intersubjectivity: an ethic that embraces and strives toward a careful (as reverent as it is agonistic) relation with otherness. Such an ethic would recognize as most fruitful the protection of the interrogative edge itself. In this sense, what Adorno seeks to acknowledge and cultivate is not simply an aesthetic sensibility (especially not one as subjectivistically defined as Habermas's often is) with respect to the "extralinguistic within" and the nonhuman. Moreover, it is what might be called an ethical-aesthetic sensibility, insofar as an ethical engagement of receptive generosity with otherness informs the aesthetic sensibilith which in turn informs the ethical engagement.

Thus far I have suggested that Adorno's sense of, and call to, engagement with extralinguistic nonidentity is more plausible and compelling than both the radical linguistic turn regarding inner nature and the instrumental-objectivating position regarding outer nature, taken by Habermas. In order to get a clearer sense of what is at stake in their differences, we must deepen the critical perspective that Adorno's work affords vis-à-vis Habermas's "turns." From the perspective that Adorno affords, Habermas's turns can and must be understood in terms of their roots in a misunderstanding of the human condition—a misunderstanding which engenders a systematic blindness to questions of nonidentity. For Habermas, as we have seen, the feature that defines the human condition, so thoroughly that we can root the transcendental conditions and norms of possible experience and knowledge in its structures, consists of a relentless *pressure* which is real and experienced in our relations with the nonhuman world and with one another. Regarding the nonhuman, this pressure is articulated in a narrative of "scarcity" through which the ubiquitous "constraint of external nature persists" (*KHI*, 58).[17]

This situation makes labor a "perpetual natural necessity," that activity to which we are "bound" in order to "keep alive," to such an extent that Habermas can speak of this instrumental action as "*the* invariant relation of the species to its natural environment." If conditions of labor "*bind* our knowledge of nature to the interest of possible technical control over natural processes" (as opposed, say, to technical control being one among several possibilities and purposes of knowledge regarding the nonhuman), that is because the pressure of scarcity is weighty and relentless enough in our natural history to bind our "deep-seated structures" to our overriding efforts to wrestle with scarcity (*KHI*, 35, my emphasis). Almost like metamorphic rocks, our structures of acting and knowing are compacted and crystallized in their very being under this pressure. Or are they like diaphragms whose structure and muscular activity develop according to the requirements of a particular air pressure? Our pressure-bound activities bind our knowledge by constituting a structure that continues to function even when the pressure is released. Thus, even when science is released from the immediate demands of productivity and develops "*beyond* merely engendering technically useful knowledge" (*PDM*, 113), it remains nevertheless within the limits of an objectivating instrumental framework.

Similarly, the structures of intersubjectivity are formed under pressure. Earlier in this chapter, we saw how Habermas follows Meade in understanding the very metamorphosis of the prehuman into the human in terms of the development of a communicative action that forms under "the pressure to adapt that participants in complex interactions exert upon one another." Presumably this social pressure is substantially engendered in light of scarcity pressures. But it also seems that collective action, on Habermas's account, generates its own pressures to conform, through which the compacting and particular crystal structure or muscular functioning of communicative action *as such* develops. The significance of this structure is clear in Habermas's definition of "normal communication" as that which takes place "under pressures to decide" for purposes of action-coordination. As we have seen, these pressures constitute our communicative action in a manner *bound* by the regulative idea of a telos of consensual identity, so that we must remain within the limits of the "straight-forward, informative, relevant" and "avoid obscure, ambiguous, and prolix utterances." Just as the pressures of scarcity squeeze out any possible cognitively fruitful encounter with the nonidentity of nonhuman nature, so too social pressures smother the space for "tarrying with the negative." Of course, these pressures may be released in discursive argumentation; but even here, like an overly flexed muscle or a muscle formed through a singular kind of flexing, the structure of the pressurized "normal" defines the operation.

Now, Adorno's critique of the underlying assumptions of Habermas's transcendental pragmatics obviously cannot supplant the idea of a pressurized context with that of a pressure-free development; for the narrative Adorno constructs with Horkheimer is replete with a sense that, as the last words of the *Dialectic of Enlightenment* say, "all things that live are subject to constraint" (258). Thus—owing to our finitude and the related extent to which the world subjects us to myriad necessities, demands, contingencies, and threats—thought, "as such, before all particular contents, is an act of negation, of resistance to that which is being forced upon it; this is what thought has inherited from its archetype, the relation between labor and material" (*ND*, 19). This resistance to the immediate pressures of the world is part of the very being and functioning of conceptual thinking and connects it significantly to "the control of nature." In the violence and counterpressure lodged in thought's opening, thought is a "moment of the reality that requires [its] formation" (11). The pressures thought resists come to be lodged in its own being in the aspect of "compulsion" in its identifications. Yet, in a repeated theme, "the point at which thinking aims at its material is not solely a spiritualized control of nature. While doing violence to the object of its synthesis, our thinking heeds a potential that waits in the object, and it unconsciously obeys the idea of making amends to the pieces for what it has done. In philosophy, this unconscious tendency becomes conscious" (19). At least in negative dialectics.

I have tried to develop this insight epistemologically and ethically in Chapter 2. Here I will simply elucidate some key points of divergence, or depressurization: points where Adorno departs from the pressure narrative in ways that are central to understanding how his position affords a critique and movement beyond the Habermasian paradigm.

For Adorno, the resistant aspect of thought to the pressures of the "given" does not tend in a *singular one-dimensional* way toward the control of nature. Rather, or *most fundamentally*, as a resistance to the "given," thinking opens *beyond immediacy* toward other possibilities for perceiving, conceptualizing, and acting in the world. If this opening, this impulse to escape "being forced to bow," can be and often is articulated in terms of control, and if it is never free of a controlling transgressive aspect, in its inception it is *more indeterminate and polyvalent*. As elucidated above in Chapter 2, under the pressure of an overwhelming world the idea of *mana* emerges, through which the tree is recognized as more than the concept "tree." Indeed, "language expresses the contradiction that something is itself and . . . other than itself" (*DE*, 15). Language is both a resistance to the immediacy of the world's pressures *and* a resistance to the immediacy of its own identifications. Thus there is rarely a clear line between "the world's pressures" and the pressures stemming from our own conceptualizations.

Again, our opening can be inhabited and articulated in ways that allow and encourage selves to expand and deepen the talonlike grasp of an instrumental reason; or, it can be pervaded with seemingly impervious taboos that function in various ways to squash the potential negative dialectic and freeze in terror. But the resistance to the pressures of immediacy that is provoked by language "as such" is not reducible to these directions. As this resistance, thought is a "moment of the reality that requires [its] formation" (*ND*, 11). The pressures which that thought resists come to be lodged within its own being in the aspect of a "compulsion" and "coercion" in its identifications. As resistance to the pressures of immediacy, however, thought is in essence capable of *resisting itself* insofar as it bears, and discerns that it bears, these pressures in its own movement. In other words, intrinsic to thinking's movement (when it truly is *thinking*) is an impulse toward depressurization that draws thought into a practice of self-critique and opening beyond its own compulsion. This self-critique and opening of thought bears compulsion and pressure within itself in ways that are irreducible to such compulsion. This is part of what Adorno has in mind when he writes in *Negative Dialectics* that "the force of consciousness extends to the delusion of consciousness" (148).

Thought, then, is not singularly *bound* to a pressurized instrumental framework in response to the world's pressures, nor is it *bound* to articulate itself exclusively in terms of social "pressures to decide"; rather, in its very being, thought opens onto another possibility, and there the grip of the pressure narrative begins to loosen. As we develop cognitive and practical capabilities that reduce our fear on a material level, and as we come to recognize the extent to which our own existential, cognitive, and institutional constitution of otherness plays a major role in engendering the very pressures to which we thought we were merely responding, we can begin to move beyond simply resisting pressures in instrumental or assimilating ways: we can move in directions that articulate and practice greater possibilities for freedom, mutual thriving, giving and receiving. It is possible to move toward higher possibilities of a "togetherness of diversity" in which "nonidentity is the secret *telos* of our identifications" and is the form of thought's resistance to the pressures of immediacy (*ND*, 149). In sum, from the vantage point of negative dialectics, the pressure thesis concerning our relation to the nonhuman and social worlds ceases to be a universal metanarrative and becomes one (albeit important) aspect of these relations which varies historically in its relative salience. The twin sovereignties of instrumental mastery and exclusive intersubjectivity can be dethroned.

Unless. Unless, as Adorno thinks has most often been the case, we treat pressure as singularly ontological and make *normative* our pressurized responses. Adorno opposes this "spirit of the age" every bit as persistently as Habermas seems to affirm it. In Habermas's "transcendental pragmatics," Adorno would identify a reification, a cultivation, a mobilization of pressure that, in spite of itself, puts the squeeze on nonidentity. "What is called 'communication' today is the adaptation of spirit to useful aims," to an exclusive consensuality born in pressures to decide and coordinate that leave little room for a "lingering eye" or a voice that seeks to discern nonidentity. In a passage worth repeating, Adorno's critique has an uncanny connection to the Habermas passages under discussion: "The straight line is now regarded as the shortest distance between two people, as if they were points. Just as nowadays house-walls are cast in one piece, so the mortar between people is replaced by the pressure holding them together. Anything different is simply no longer understood" (*MM*, 41).

Of course, I am giving an overly "pressurized" reading of Habermas. Clearly there are, in addition to the provocative openings I have noted in my elucidation of his position, essays and sections of Habermas's writing that spill far beyond the confines of an insistently consensualist and exclusively linguistic intersubjectivity. There is the suggestive essay on Walter Benjamin which contemplates the tragic possibility of justice without happiness.[18] There are moments when Habermas seems to expand his aesthetic account in ways that might open his framework to a much greater engagement with nonidentity. The "decentering" provoked by avant-garde art, he writes, "indicates an increased sensitivity to what remains unassimilated in the interpretive achievements of pragmatic, epistemic, and moral mastery of the demands and challenges of everyday situations; it effects an openness to the expurgated elements of the unconscious, the fantastic, and the mad, the material and the bodily—thus to everything in our speechless contact with reality which is so fleeting, so contingent, so immediate, so individualized, simultaneously so far and so near that it escapes our normal categorical grasp" ("QC," 201). Yet these moments are rare in his massive corpus. And they are never allowed to interrogate the dominant framework. Frequently, as is true with respect to this last passage, they are denied epistemic and moral relevance; or they are assigned to marginalized and subordinate categories like that of the "world-disclosive." But why would our increased openness and sensitivity to the unassimilated not be of *great* epistemic and moral import? Primarily, I think, because of the centrality of the pressure narrative for Habermas's understanding of the human condition and his corresponding ontologization and normalization of instrumental and intersubjective logics of identity and consensuality. Thus Habermas is led to a pressurized reading of *himself*—the lines of which I have attempted to discern.[19]

Yet "what difference does this difference make? I think the effects might be far-ranging. I conclude the present chapter by illuminating some of these differences in terms of epistemological and ethical "modes of conduct," leaving a discussion of the possible political divergences for the next chapter. Chapter 2 above represents a longer response to the question posed, so I will limit myself here to exploring the significance of Adorno's juxtaposition of agonism and consensus with respect to questions of epistemology, ethics, and individualization.

If selves confront significant and ineliminable extralinguistic (as well as linguistic) nonidentity, and if the pressures of existence ought not be so exhaustive of our account of the human condition and our accompanying norms, then the *telos* of consensuality can and ought to be dislodged from its singularly privileged position in our theoretical and ethical efforts. The inexhaustibility of nonidentity implies that no identification or consensus will be adequate. It further implies that a significant aspect of giving and receiving between nonidentical beings will involve precisely the ways in which they contest each other's self-enclosures. From this perspective, a discordant agonistic impulse has not simply a temporary or subordinate value but one that is permanently inscribed in our existence. If human "life that lives" is rooted in practices of receptive generosity among nonidentities, then one regulative ideal which ought always pull on us is that of resisting not only extant consensuses but also the insistence that our very resistance be *exclusively* animated by the regulative ideal of consensus.

As we have seen, Adorno, far from simply opposing consensuality, carefully formulates and cultivates an important point of consensual striving in his ethical constellation. Yet he locates our highest possibilities in a "morality of thinking" that situates itself in the tension *between* these discrepant (but related) pulls toward consensuality and a more agonistic dissent. Here we are more likely to escape our somnambulistic tendencies and engage the world in a more receptive and generous manner. Outside such a tension, we are most likely to reify, undermine, and be insufficiently animated by the ideals that guide us. For Adorno, it is not enough to make space for nonidentity, as Habermas does, simply in the recognition of a factual reemergence of different perspectives in open communicative action—no more than it would be sufficient to make space for consensuality merely in recognition of our more "herdlike" qualities. Rather, both concerns, so vital to our engagements and so often dormant, require careful articulation, solicitation, and juxtaposition.

Significantly, a "morality of thinking" animated by agonistic constellations is able to resist the demand for a singular consensuality without succumbing to the charge of committing a "performative contradiction" of the sort that Habermas generally levels against those who would depart from "communicative ethics." Adorno never *denies* the crucial role of the ideal of consensuality in dialogue but, rather, supplements and contests it with the competing pull of dissent and ambiguous lingering. These points will wax and wane with respect to each other in our engagements; by virtue of their internal relations and respective significance, however, neither pull is ever entirely absent, and both ought to be vibrantly present over the larger course of our dialogues. From this perspective, the charges might indeed be reversed (somewhat tongue in cheek) such that the singular ideal of consensuality, in the absence of dissensuality and ambiguous lingering, involves—if not a performative contradiction—a provocative contradiction insofar as it conceals the illusive and damaging aspects of its own telos, thereby undermining the possibility and likelihood of articulating knowledges and situations more worthy of our consent.

The "modes of conduct" animated by Adorno's agonistic constellations can be further articulated by briefly contrasting them with Habermas's depiction, discussed earlier in the present chapter, of the types of consensually driven action: theoretical discourse, practical discourse, and the formation of personal identity. Concerning the first, for Adorno, the project of gaining empirical and theoretical knowledge significantly involves our efforts to engage unassimilated nonidentity—precisely the kind of increased "sensitivity" that Habermas, at his best, recognizes only in an epistemologically insignificant and rather marginalized aesthetic dimension. For Adorno, this effort may be directly cognitive or may be pursued more aesthetically in different ways. In either case, it is a vital part of our efforts to understand the world. Hence, the activity of discourse aimed at understanding involves a degree of *resistance to* (as well a need to pay a certain heed to) a "concern to give one's contribution an informative shape, to say what is relevant, to be straight-forward and to avoid obscure, ambiguous, and prolix utterances." The informative, relevant, straightforward, and unambiguous—for all their merits—are always significantly engendered by what for Adorno are always partly damaging identities belonging to what Habermas calls the "intramundane." Of course, Habermas does not affirm a positivism of the intramundane, the identified world, but rather endorses "testing processes of intramundane practice" which are governed by the *ideal* of consensus that transcends all local constraints. At question, though, is the extent of the transcending force of this intramundane testing and its singular demand that we communicate in a straightforward manner within the bounds of a pressurized problem-solving which aims at "mastering problems posed by the world"—a world which itself is often poorly identified through pressurized discourses and practices.

For Adorno, while important work can and must be done at this level, to make it normatively binding in a singular way is to demand a degree of continuity—even conformity—in thinking that weakens thought's potential capacity to break with extant identities in desirable ways. An invaluable aspect of thinking has to do with the effort to problematize the "problems," their "logics," and the identities that sustain them. Adorno seeks to solicit and animate this effort by repeatedly illustrating, affirming, and trying to make space for thoughts whose worthiness lies precisely in their "distance from the *familiar*" identities, logics, problems, relevance, and so on. The act of thinking and judging requires concerns that are more agonistic and also more attuned to the potential fruits of indefinite "lingering." The activity and power of the "world-disclosive" are not "tamed" (nor should they be tamed in any singular way). They should not have to "prove their worth" *in a one-directional manner* according to the privileged established logics of intramundane testing. Rather, thinking involves sustaining a tension of *mutual* interrogation between "intramundane processes" and "world-disclosive" paradoxical engagements with the unassimilated.

Themes of relevance for questions of practical moral discourse are not difficult to discern in Adorno's thoughts on the activity of theoretical discourse. The dialogical situations toward which one might aim (the extent and types of space, the solicitations animating such spaces, the ethical-educational-cultural projects that seem most suitable for engendering dialogical selves and communities, etc.) in order to engender a more intelligent, free, just, and thriving society, while not lacking consensuality, are characterized by an embrace of a much greater degree of agonism, lingering, and indeterminacy than one finds in Habermas's ideal discursive situation. While the latter situation certainly "tolerates" agonism, the idealizing suppositions of Habermasian discourse exert a persistent pressure in the other direction. Yet this relatively unidirectional pressure slackens the tensions that tend to animate our resistance to the dangerous aspects of our identifications. With this weakening of the "morality of thinking," we are more prone to proliferate—or be victims of—assimilation and imperialism.

However, Adorno's greater embrace of agonism does not lead to an abstract negation of Habermas's universalist rule of argumentation or his principle of discourse ethics, but rather to their repositioning. It is difficult to imagine a form of universalist respect in which "a principle that constrains *all* affected to adopt the perspectives of *all others* in the balancing of interests" does not play an important role ("DE," 65). Still, from Adorno's perspective, one can also imagine the importance of soliciting an even more receptive effort (e.g., *vis-à-vis* possible aspects of the other and its relations that the other does not acknowledge) as well as a greater degree of recalcitrance to this "universal exchange of roles" (e.g., moments of powerful insistence on the value of something in one's own or another perspective unrecognized by the other). Entwined with this receptivity and recalcitrance is a sense that receptive generosity toward otherness always takes our efforts to explore possibilities of giving and pulls them *beyond* the extant laws and norms, insofar as these laws and norms will always to some extent fall short.[20] Of course, the moment of recalcitrance in tension with any principle of universalization would find itself interrogated by that principle as well. Adorno would reject not the idea that a process of free interactive participation is a vital aspect of the legitimation of democratic norms, laws, and policy, but rather the overly domesticated process of participation that Habermas often seems to support.

In all of this, it is likely that the scope of diversity which is significantly engaged with respect and receptive generosity would be greater in the negative-dialectical ethical mode of conduct. So too would the— at once agitated and agitating—sense of irony and tragedy that would accompany many of our articulations of the principle of universalization. This sense in turn might lead to a greater caution and sensitivity with respect to whatever remains nonidentical to the orders and identities we formulate; and we might then impose those orders and identities less expansively, with a lighter touch and more "slack."[21]

Let us turn now to the question of the development of individual identity. We have seen Habermas's account of the genesis of the individuated autonomous subject in the context of consensually driven communicative action, and we can now anticipate a repositioning similar to those I suggested with respect to epistemological and ethical modes of conduct. For Habermas, the self "forms," in the sense of developing accountability, responsibility, and irreplaceable individuation, in the shifting of perspectives of communicative relations. Consensually driven intersubjectivity "forces" selves to act under the singular presupposition of responsibility as the "demands" of others pressure us into this self-relation. Yet the self's genesis in, and dependency on these demands and recognitions through which we secure our identities is not (as illustrated above) simply or ultimately rooted in extant others and communities. Rather, as the idealizing structure of consensual intersubjectivity establishes a substantial independence from concrete lifeworlds, both the demand for accountability and the need for recognition come to be located in an *anticipated* community. Before this "wider commonwealth of rational beings" whose idealized demands and recognitions we deem to be legitimate, we develop and "prove" ourselves in a self-reflection aimed at ever deeper and broader continuity and consistency.

Yet it is possible to discern here, in addition to much that is admirable, still another line that secures the pressure narrative. The pressure to decide gives rise to singularly consensual communication, which in turn pressurizes our identities under the tight grip of an unrelenting consistency—in the midst of which the very existence of pressure becomes increasingly indiscernible. Of course, a self that did *not* feel the pull before others to account for its ideas, defend and develop its arguments, wrestle with inconsistencies, and develop in this process would be a lousy partner in dialogue, would probably have scant recognizable intelligence, and would likely be dangerous. But the singularity of this pull also has accompanying dangers. Even under the most idealized suppositions, will not a self which is singularly animated by the imperative to be consistent tend to conceal (or fail to acknowledge and engage) aspects of nonidentity within itself, others, and the world that might fundamentally disrupt (in ways that might never be wholly recoverable within any seamless order) its identity? What if there is much in human beings and the nonhuman world that requires substantial moments of contradiction, ambiguity, paradox, and discrepancy in order to receive any illumination whatsoever? If a significant part of the human condition is this way, might not a more desirable process and goal of individual cultivation be found in a self that strives to fashion its life in the tension between the demands described by Habermas, on the one hand, and a different responsibility to nonidentity that draws the self toward explorations of inconsistency and discontinuity, on the other? Such a self might be less likely to smother the "chaos needed for the shooting star" and more likely to reveal the problematic limits of its own consistencies as a way of opening a deeper engagement with others.

No one illuminates and celebrates the virtues of inconsistency more powerfully than Ralph Waldo Emerson and Friedrich Nietzsche. Adorno suggestively constructs a constellation whereby selves are pulled by, and develop through, the demands for consistency *and* the demands to resist it. Similarly, concerning recognition, Adorno conceives of selves that would be pulled simultaneously by the need to be recognized by an anticipated community *and* by the need to resist any imaginable

community (insofar as no community achieves assimilation and reciprocal recognition without damage). Generosity toward nonidentity involves the effort to include and be included by otherness in a dialogical
recognition; but it also involves the effort to transcend and break free
of the undesirable aspects and limits of all inclusion.

Post-Secular *Caritas*, Coalition Politics, and Political Economy

My reading of Adorno has moved us a substantial—and I think desirable—distance from Kant's narrative of self-given sovereignty. Traveling in a direction Kant began to elaborate in the *Critique of Judgment*, I read Adorno as developing a dialogical ethic of receptive generosity animated by a constellation of agonistic solicitations. This articulation of ethics advances beyond understandings of morality based on universal commands or categorical imperatives that could be given by even the most "unpracticed understanding." The respect suggested in such a morality is significantly undermined by the sovereignty narrative which grounds it, so that, in the absence of receptivity, "respect" may beget an oblivious violence toward others exemplified by Kant's complicitous relationship with the "murderer at the door." From the early 1930s until his death, Adorno criticized the oblivion, assimilation, and imperialism engendered by such self-given sovereign respect. Yet, reformulating morality involves more than constituting another set of moral declarations. Not that there are no universal commands which ought to compel us in a rather simple way. There are some. The larger and deeper task of living ethically, however, is articulated through infinitely complex dialogical relations requiring an agonistic receptive generosity that exceeds any law or sum of declarations. Thus, Adorno's ethics locates itself in the midst of numerous conflicts and paradoxes in an effort to stir thought from its reified obliviousness with respect to otherness.

Below I shall develop possibilities, opened by Adorno's "morality of thinking," for reflecting on the relations between ethics, civil society, social movements and coalition politics, and political economy. I begin with a short discussion of Marx, but not simply because Adorno and Habermas can be read as discrepant responses to this revolutionary who inspired them, frightened them, and sometimes just came up short on crucial questions. More important, Marx allows us to situate the philosophical issues in a context of political, social, and economic concerns. Next, I shift to Ernesto Laclau and Chantal Mouffe, who probe beyond Marx with a "radical democratic liberalism." In light of the ethical and political problems in their work, illuminated by Bernice Johnson Reagon's discussion of coalition politics, I reopen the possibilities in Habermas and Adorno for addressing problems that are insufficiently dealt with by Marx and radical liberalism.

Communist Time: Marx and Receptive Generosity

"Marx," Kolakowski notes in the opening line of his four-volume work on Marxism, "was a German thinker."[1] An exhaustive account of the meaning of this statement would require a careful study of Marx's relation to Kant, Fichte, Schelling, Schiller, Hegel, as well as the "Old" and "Young" Hegelians that is beyond the scope of this book.[2] Still, in order to see how the themes of giving and receiving get transfigured in Marx's project, it is important to situate Marx in relation to my reading of Kant. The transfigurations are not just Marx's own doing, but my only concern for now is their articulation within Marx's texts.

The early Marx understood his project to be oriented by the "Copernican revolution" in philosophy. The critique of religion, he wrote in his "Contribution to the Critique of Hegel's *Philosophy of Right*: Introduction," culminates when "[man] will revolve about himself as his own true sun."[3] For Kant, the Copernican revolution in philosophy meant, in contrast to metaphysics, that just as Copernicus dared "to seek the observed movements, not in the heavenly bodies, but in the spectator," so ought we understand nature in terms of our own spontaneous categorical synthesis.[4] Thus, from the idea that we revolve about the sun, the subject in effect becomes the sun, giving order and experienceability to the world. For Marx (and Kant), the metaphysical illusions that kept us from revolving about ourselves functioned in practice to

keep us in chains—and Marx understood the task of freedom in revolu-
tionary-historical terms.

Man's revolving about himself has two related senses for Marx. At
the level of the species, it means that man will "adopt the species as his
object" of conscious historical self-development, including the earth,
which is our "inorganic body" (*1844*, in *MER*, 75). Here, the human
species, as a collective subject, freely gives to itself and the world the
form and (largely) the content of its being. No longer awed by a social
or natural world that stands over people, communism thus "treats all
natural premises as the creatures of hitherto existing men, strips them
of their natural character and subjugates them to the power of united
individuals . . . rendering it impossible that anything should exist independently of individuals" (*German Ideology*, in *MER*, 193). No one has
developed this theme of self-giving collective subjectivity more powerfully than Georg Lukács. The community of revolutionaries, by taking
control of the conditions of existence, effectively becomes the synthetic
condition of possibility for being and knowing.

Entwined with this collective solarity is a diversely articulated individual solarity. "It is as individuals" that revolutionaries participate in
the metahistorical project (*German Ideology*, in *MER*, 197–98). No
longer "levelled" or "average," nor "envious" and "destructive" of "everything [including talent] which is not capable of being possessed by
all" (*1844*, in *MER*, 82–83), the selves of communist society are to be
"free" and "conscious," interacting in community so that only "with
others has each individual the means of cultivating his *gifts* in all directions" (*German Ideology*, in *MER*, 197, my emphasis). Yet with this pluralization of solarity, the principle of illumination has clearly changed.
We have moved beyond metahistorical reflections on a sovereign collective subjectivity, and we now find ourselves closer to the most exuberant passages in Kant's reflections on aesthetic ideas concerning a
flowering of consciousness through giving and receiving. But now we
see a flowering *being*, flowering *history*, and gone are the scattered reflections on finitude, failing, violence, and unresolved tensions. Communism, after all, is the *resolution* of all riddles, and it knows it. It is
more beautiful than sublime.

The new principle of shining and of cultivating one's gifts is rooted
in a receptive appropriation (no longer narrow, possessive, and individ-
ualistic) through which we are multiplicitously nourished and formed—
like the five senses—by "the entire history of the world down to the

present," by all others with whom one is united, and by "all nature" (*1844*, in *MER*, 89 and 75). And this receptive cultivation of each person's gifts in turn has receptive others deeply inscribed in its own being: gift cultivation is in its essence *generosity*. Beyond the narrow logic of exchange and "bourgeois right," in some distant future "society inscribes on its banner: From each according to his ability, to each according to his needs" ("Critique of the Gotha Program," in *MER*, 531).

Marx does not compromise this exuberant, unrestricted giving and receiving. Marx understands previous time as a semicongealed river, not frozen solid but thick with ice and slush: barely flowing, flowing unevenly, clogged and flowing backward in spots. We are in "prehistory" when the frozen objectified labor of the past subjugatively structures present living labor. Indeed, "The tradition of all the dead generations weighs like a nightmare on the brain of the living" (*Eighteenth Brumaire*, in *MER*, 595). A nightmare? In the most terrifying recesses of this restless sleep, the nightmare concerns time itself. Time now is the *stealing* of time, it is in two senses anti-time: time is robbed of the generous living relations that are essential to time as the movement of freedom, consciousness, and cultivation; and time is structured such that the past, congealed in capital and class relations, steals time, takes away from the living present its free time, binds it in "forced labor" and in enslaving consumption (*1844*, in *MER*, 74 and 93).[5]

Marx imagined communism temporally as a kind of thaw, a vernal awakening. The present would be released from the past's freezing grip and begin flowing in accordance with its own unrestricted dynamism. Thus "the proletariat rids itself of everything that still clings to it from its previous position in society"; it becomes free, universal self-activity (*German Ideology*, in *MER*, 192). It is not so much that the present rips itself out of the thieving anti-time of prehistory. Rather, the present's giving and receiving develops a power capable of thawing the past—by means of a sovereign receptivity which transforms prehistory in spite of itself into giving-time. "Communism differs from all previous movements in that it overturns the basis of all earlier relations of production and intercourse, and for the first time consciously treats all natural premises as the creatures of hitherto existing men, strips them of their natural character and subjugates them to the power of the united individuals . . . [and] treats the conditions created up to now by production and intercourse as inorganic conditions, without, however, imagining that it was the plan or the destiny of previous generations to give them material" (*German Ideology*, in *MER*, 193–94). Communism is the generous and therefore "dynamic principle of the immediate future" (*1844*, in *MER*, 93), for it receptively brings the past under the dynamism of the present.

But what does the time of communism give? It gives *time*: freer and free time. By making possible "unrestricted self-activity," communism frees human temporality so that in the living present each individual— through an unrestricted reception of all past moments, beyond all "fixation of social activity"—gives in such a way that "hunter, fisherman, shepherd, [and] critic" flow together in a person's day in a rich cultivation of gifts (*German Ideology*, in *MER*, 160). While earlier Marx had hinted that the absolute freeing of time might permeate all activity, he later restricts himself to more moderate claims, distinguishing between a "realm of necessity" (material production) and a "realm of freedom." In the former realm, the possibilities for giving and receiving are substantial but are nevertheless limited by "mundane considerations" and an aspect of heteronomy. Thus, communist production, beyond the task of organizing itself "under conditions most favorable to, and worthy of . . . human nature," has the project of giving time, giving free time beyond the realm of necessity. "Beyond [necessity] begins that development of human energy which is an end in itself, the true realm of freedom, which, however, can blossom forth only with the realm of necessity as its basis. The shortening of the working day is its basic prerequisite" (*Capital*, vol. 3, in *MER*, 441).

Free time is where time and beings are utterly drained of opacity; where "man is not lost in his object . . . [because] the object becomes for him a *human* object or objective man"; where "the objective world becomes everywhere for man . . . the world of his essential powers . . . the objectification of himself, [and where that world] becomes objects which confirm and realize his individuality" (*1844*, in *MER*, 88).

If the manifold relations among individuals are the condition for the collective solar subject, it is in turn the movement of the whole, the "united individuals," which makes possible the dynamic relations of giving and receiving, by properly guiding economic, social, and political development. Prehistory and revolutionary activity are to combine so that the dynamic between the vast majority of individuals and the collective voice of unified subjectivity attains a grand fluidity that requires neither a state nor a separate communist party—Marx strongly warns against both—but simply the articulations of communists who are "practically and theoretically the most advanced . . . [who] always and everywhere represent the interests of the movement as a whole . . . [and who] clearly understand the line of march" (*Communist Manifesto*, in *MER*, 484). For Marx, this is to be a radically participatory democratic project, like the one described in his essay that ironically heroizes the Paris Commune revolt. Beyond all the specific actions taken by the communards, "the great social measure of the Commune was its own working existence. Its special measures could but betoken the tendency of a government of the people by the people" ("Civil War in France," in *MER*, 639). They "could but," but . . .[6]

I will not recount here the dangers of this vision when it confronts the recalcitrance of history: the way in which all opacity and resistance is converted into an evil to be driven from time; the way in which the dream of a collective subject combined with a certain epistemological privilege tends toward the coercive party structure that Marx explicitly sought to avoid; the conquest of the earth and the disciplinary mobilization of human beings. The beautiful and manifold giving-and-receiving, which was to constitute a species-subject, itself *required* this collective subject to make us easily receivable to one another. This subject was in some ways a materialist version of a God whose power was born and contained in *caritas*. But like Augustine's God, the collective subject was often a jealous God; and jealous gods, as Nietzsche tells us, get ugly.

Marxists, neo-Marxists, post-Marxists, and progressives more generally have sought to formulate ethical and political positions that provide a sharp critical edge vis-à-vis contemporary capitalist societies and that offer more desirable historical possibilities while simultaneously trying to avoid the terrors of twentieth-century communist experiments. In this context, we shall explore further the ethical and political possibilities of Adorno and Habermas. I think receptive generosity must play a vital part in reformulating a progressive politics which moves beyond Marx to address contemporary societies in an effective and desirable manner. In contrast, one major effort to reformulate a "post-Marxist" political theory abandons receptive generosity almost entirely in favor of a radical liberalism based on liberty and equality. So, before discussing Habermas and Adorno, let's consider the work of Laclau and Mouffe, who offer an alternative to both Marxism and mainstream liberalism.

Left without Generosity

Laclau and Mouffe are disillusioned both with the politics of sovereign collective subjectivity and with Marxism's weakness in engaging the politics of contemporary social movements. In contrast to Marxism's frequent understanding of coalition politics as a defective historical exigency to be dissolved in a unified, harmonious, and transparent collective subject, Laclau and Mouffe affirm a contestational politics of diverse coalitional movements as historically more fundamental, ethically more desirable, and politically more tenable. They draw support from neo-Nietzschean reflections on radical contingency, multiplicity, and antagonism in history which are at odds with many of Marx's assumptions. Ethically, they affirm a "radical liberalism" which seeks to spread equality and liberty in radical democratic fashion.[7] But while liberty and equality are transfigured in ways that turn up the historical volume, Laclau and Mouffe seem to flip off the switch when they come *caritas*.

Laclau and Mouffe argue for a "new logic of the social" that might hegemonize (gather together and transfigure) new social movements around an agonistic embrace of the norms of liberty and equality which have animated historical movements for more than two hundred years. Their sense that these norms are sufficient is exemplified when Mouffe writes that "the problem . . . is not the ideals of modern democracy, but the fact that its political principles are a long way from being implemented."[8] Ethically and practically, their ontology deconstructs all essentialist efforts to restrain the proliferation of sites wherein the ideals of equality and liberty find expression. But it goes no further.

The *status* of these ideals of equality and liberty, however, is transformed when Laclau and Mouffe reject such metaphysical foundations as transparent self-grounding subjectivity, reason, and history. Henceforth, they argue, these values must be understood—and embraced with a tragic heroism—as utterly *"contingent* historical projects," products of struggles that have generated the particular traditions which precariously make and unmake our identities. Laclau writes: "We happen to believe in those values."[9] To those who rejoin that this statement is unlikely to convince anyone who does not already "happen to believe," Laclau responds with a skepticism toward both the persuasive efficacy of rationalist arguments and the rhetoric of persuasion itself, insofar as "persuasion . . . structurally involves force" and "persuasion is one form of force."[10] Henceforth, ethics is subsumed within political struggles that seek a hegemony involving "the construction of a new 'common sense' which changes the identity of the different groups, in such a way that the demands of each group are articulated equivalentially with those of others"[11] around the ideals of liberty and equality.

The change in status of these ideals, their subsumption within political struggles, is not without effect on their substantive meaning, function, and articulation. The erosion of foundations, "far from being a negative phenomenon, represents an enormous amplification of the content and operability of the values of modernity."[12] Laclau and Mouffe elucidate this point by distinguishing between a *foundation*, which has an internally "determining and delimiting" relation to what it founds, fixing and constraining the content and function of the founded, and a *horizon*, which is "an empty locus" with an essentially "open-ended" character. A horizon is a soliciting yet essentially inexhaustible reference by means of which a group "constitutes itself as a unity only as it delimits itself from that which it negates."[13] Thus, groups can gather themselves around the horizons of equality and liberty through numerous open-ended struggles against inequality and subjugation in which the meanings of equality and liberty take form as developing "social logics."[14] By appropriating equality and liberty as social logics, Laclau and Mouffe hope to intensify the best aspects of the Enlightenment.

Equality and liberty are understood, by Laclau and Mouffe, to reciprocally define and reinforce each other. Equality is significantly an equality of liberty, and liberty has an egalitarian structure. Yet insofar as their ontology precludes essentialist understandings of these values, and views them as correlates of an antagonistic ensemble of social forces, equality must mean more than just equality of liberty—and this "more" disrupts the reciprocally supportive relationship between the two values. For the very identity and existence of equality and liberty hinges on a certain *solidarity* with the *common* project of radical democracy through which these values can be maintained and can proliferate. Thus equality—more thickly interpreted as identifying with *the same* project, the "chain of democratic equivalence"—is the condition for constituting the "we" by which equality of liberty can exist. Granted, the "chain of democratic equivalence" necessary for hegemony is bound to this pair of very *formal* and negatively defined values, rather than to substantive notions of the good life. Yet, insofar as this equivalence is the condition for egalitarian spaces of difference, difference and liberty are subordinated to equality, approximating the very type of systemic totality they seek to avoid. The dangerous aspect of this idea is striking in Mouffe's claim that "to construct a 'we,' it must be distinguished from a 'they' and that means establishing a frontier, defining an 'enemy.' . . . [C]onsensus is by necessity based upon acts of exclusion."[15]

Aware of these concerns, yet insisting that we can never entirely escape the problem, Laclau and Mouffe argue that the totalizing requirements of public life (equality as equivalential chain) can be juxtaposed with other values (liberty as autonomy) to check totalization and generate multiplicitous emancipations. Hence it is not simply the reciprocally supportive relation between equality and liberty, but also their *tensional* juxtaposition, their reciprocally limiting and antagonistic relations, that gives life to radical and plural democracy.[16] Such agonistic juxtapositioning of values has affinities with Adorno's ethical constellation.

The freedom and power of diverse social movements almost always hinges on their ability to combine with other struggles, and Laclau and Mouffe view radical democracy to be precisely the identity-modifying focus around which the chains of equivalence necessary for such coalitions can be established. Yet given their ontology of difference, antagonism, and contingency, "this total equivalence never exists: every equivalence is penetrated by a constitutive precariousness." Given that total equivalence is always horizontal, all hegemonic projects established on totalizing claims will be false, and it is likely that various social groups will find such projects transgressive of aspects of their identities, problems, and aspirations. A movement which at once recognizes this irreducible condition and embraces the ideal of egalitarian liberty will allow and encourage the logic of autonomy both to transfigure and limit the logic of hegemonic equivalence. "To this extent, the precariousness of every equivalence demands that it be complemented/limited by the logic of autonomy. . . . [T]he demand for *equality* is not sufficient, but needs to be balanced by the demand for *liberty* . . . the irreducible moment of the plurality of spaces."[17] This plurality of spaces makes possible an open-ended renegotiation of the terms of radical democratic hegemony: equality and liberty. This is partly a strategic move, insofar as Laclau and Mouffe believe an open and plural ideal can unite diverse movements in collective action. But it is also an ethical move, based on the idea that the horizonal character of their ideals solicits a deepening and broadening of ethics that requires plural reformulations.

In short, Laclau and Mouffe seek to reformulate an indeterminate and open hegemony. "Radical democracy makes this openness and incompletion the very horizon on which all social identity is constituted."[18] And "the fullness of the social . . . manifests itself . . . in the possibility of representing its radical indeterminacy."[19] What this reformulation means politically, as the project of radical democracy multiplies spaces, diversifies its struggles, solicits subjugated voices, and gathers them in a growing hegemonic formation, is that "through the irreducible character of this diversity and plurality, society constructs the image and the management of its own impossibility."[20] Workers, blacks, women, ecologists, gays, lesbians, consumers, anti-imperialists, and others are transfigured through the practice of radical democracy and participate in a coalition that simultaneously draws them toward a new "common sense" and guarantees autonomous spaces for contestation, reformulation, and marking specificity. In the process, a precarious and renegotiable balancing act between identity and difference is instantiated— one that offers, Laclau and Mouffe claim, the greatest possibilities for human emancipation.

But does this position push as high as it ought to, given Leclau and Mouffe's philosophical embrace of contingency, opacity, and agonistic entwinement of differences? And does their construal of equality and liberty provide an ethical standpoint sufficient for the coalition politics they seek to embrace? How are we to imagine this democratic gathering of diverse groups, this community of impossibility, these relations between selves and others? How are we to characterize ethically the exchanges and movements between people? What animates them?

As is evident, Laclau and Mouffe offer two indeterminate logics and, finally, a transfiguring mixture of them. The first is a logic of hegemony, equivalence, and equality, where "being with" others is imagined as a movement in the direction of subsumption within a singular identity. Difference is not necessarily eradicated here, but is transfiguratively assimilated within a totality. Human relations are imagined as movements aimed at seducing-into-the-whole. The animating principle of these exchanges is a desire to be with the other as a part of the common. The second logic is that of autonomy, plurality, and liberty. Here, "being with" others is imagined as a movement in the direction of absolute difference, where each group (or, more radically, each self, each subself . . .) would tend toward the horizon of "auto-constitutivity," incommensurability, absolutely particular identities that would be "unable to communicate with each other."[21] Envisioned here is a retraction and dissolution of exchange, a deepening vacuumlike void of impassability. The animating desire: undisturbed transparent atomism.

Of course, Laclau and Mouffe dismiss both modes when conceived of as foundations or achievable endpoints. They embrace these modes only as indeterminate and antagonistic horizontal logics—mutually transfiguring, reciprocally limiting. In a precarious middle ground, these logics are to do battle in a manner most conducive to openness and to a freedom which escapes the tyranny of identity, whether that of the whole or that of the part.

Laclau and Mouffe are correct to identify the importance of equality and liberty for progressive coalition politics. Furthermore, their reconstrual of these principles as horizontal social logics radicalizes them in a manner consonant with Adorno's "morality of thinking." Their tensional juxtaposition illuminates and might help check the dangers that accompany a radicalized liberalism. What is lacking in their formulation, however, is an ethical account of the possibilities and heights of being with others *as others*: striving to engage their otherness. Coexistence constantly disappears into the singularity of the whole or the part. Mutual limitation is to prevent these logics from accomplishing total disappearance. Yet this opening which forms in the incompleteness of disappearance does not provide an adequate ethical account of "being with" others—the possible agonizing grandness of plurality. Lacking an ethic that solicits a more receptive and generous effort to engage otherness, might we not simply oscillate between relations of assimilation and indifference?

An ethic of receptive generosity should not *replace* the twin logics offered by Laclau and Mouffe. However, a solicitation of the generous and receptive desire for the other *as other* must enter into this constellation so that equivalence and autonomy are drawn into an imagination of community animated by a desire for others' otherness, with all the cooperation and agonism this implies.

Laclau and Mouffe are correct that we must abandon a politics of collective subjectivity. And they are right that a central aspect of progressive politics must articulate itself through coalitions of diverse social movements. Any effort to sketch a possible political direction from the ethics of receptive generosity must be able to address fruitfully the questions that arise on this terrain. No one has addressed these questions more profoundly than Bernice Johnson Reagon, who has been active in coalition politics for more than three decades. After exploring the limits of Laclau, Mouffe, and Habermas in light of Reagon's questions, I will further sketch a possible ethico-political horizon onto which Adorno's thinking might open.

Reagon and the Agonies Of Coalition Politics

Bernice Johnson Reagon's essay "Coalition Politics: Turning the Century" (hereafter "CP")[22] helps us grasp the questions, challenges, and dangers of grassroots political activity among groups of people with different backgrounds, identities, experiences, understandings, problems, relations to power, aesthetic sensibilities, and so forth. A black woman (and a member of the singing group Sweet Honey in the Rock) addressing an audience mostly of white women at a music festival significantly animated by the theme "Woman-Identified Women" (which many see realized only in lesbian relationships), Reagon discusses the difficulties of her presence at the festival—a site of coalition. For the idea that gender is *the* fundamental oppression, demanding women-only solidarity, does not resonate well with the experiences of many black women concerning sex with men, the importance of solidarities with black men, and issues of race and class. There are profound disruptions at the edges of the encounters at the festival. With playful seriousness, she says: "I feel as if I'm going to keel over any minute and die" ("CP," 356). Efforts to engage and work with others very different from oneself are usually wrought with serious difficulties, anxieties, dangers to one's identity, and so on. "That is often what it feel like if you're *really* doing coalition work. Most of the time you feel threatened to the core and if you don't, you're not really doing no coalescing" (356). Efforts to build coalitions of diverse groups are often fundamentally threatening because many of the perspectives and practices that you take to be unquestionable aspects of your identity are challenged by others who suggest that what you hold dear is in fact trivial, illusory, oppressive, obnoxious, slavelike, unhealthy. The limits and contingencies of our identities as well as the recalcitrance of others to them appear with a depth and frequency that can take our breath away. And if it doesn't, Reagon suggests, the kinds of encounters from which a rich coalition politics might develop are probably being avoided.

Because of the agonistic and agonizing character of coalition politics, "you don't go into coalition because you just *like* it" ("CP," 356). And because what people most often *do like* is the comfort of their established identities, the encounters with otherness involved in coalitioning are often obfuscated—along with the entire coalitioning effort. Engaging others in these situations involves grappling with differences by which you are at once attracted and repulsed, differences that can turn you inside out. For the "and" of "attracted and repulsed" is an extremely precarious place: bearing the burdens of this existential stress, it is easy to fall so heavily on the side of the "repulsed" that one's response hardens into "against." A variety of strategies can be deployed to avoid differences and secure one's identity. Most obvious are direct efforts to subjugate others. Or a group might declare its own struggle and identity to be the privileged location in history. Or a self or a movement (as Reagon suggests was partly the case at the festival) might have a rigidly secured identity as well as an understanding of itself as open to difference. In this case, the group opens its doors with a posture of inclusiveness; but the superficiality of the posture is often readily apparent to those who are "included." Reagon taunts: "You don't really want Black folks, you are just looking for yourself with a little color to it" (359). With luck and careful control over who gets "included," perhaps the group can recruit lots of "themselves with a little color to it," avoid the "Black folks," and maintain its posture of openness for a time, while in fact being closed, unaltered, learning little. More likely, though, "Black folks" are bound to find their way into the formerly barred room of comforting identity. Then, "the first thing that happens is that the room don't feel like the room anymore. (Laughter) And it ain't home no more" (359). At this point, one can begin the agonizing work of trying to come to terms with differences— or one can retreat into another strategy of identity securement. Because leaving "home" is painful, burnout and indifference continually threaten to provoke an apolitical retreat. Reagon fully recognizes that "you can't stay there" long in the midst of these threatening encounters. Coalition politics is at best an intermittent activity that necessitates retreat to more comfortable relations which provide types of strength and nurture mostly absent from coalition activity. Nevertheless, she enjoins us to return to this activity repeatedly, with a watchful eye against those strategies which are likely to subvert it.

In part, Reagon's position rests on a strategic sense, which she shares with Laclau and Mouffe, that peoples diversely subjugated need one another "in order to survive" ("CP," 365). But for Reagon the issue is deeper and higher than survival. It is a question of "turning the century with our principles intact" (363). "The thing that must survive you is . . . the principles that are the basis of your practice" (366). Most important to the possibility and outcome of coalition politics is an ethical survival that exceeds mere physical survival. The principles to which Reagon refers draw her vitally in the direction of encountering others *as others* and sustain her commitments even in the midst of great difficulties and dangers. The liberal ideas of equality and liberty repeatedly animate her text and life. They inform her sense of the importance of all people's struggles against subjugation as well as her efforts to let others be. Yet she radicalizes liberty and equality in a manner similar to Laclau and Mouffe, freeing those ideals from essentialism and grasping them as open-ended and expansive. Thus, referring to a song she had written years before in which she had tried to expand the content of liberty and equality, but of necessity failed to go far enough, Reagon says: "If in the future, somebody is gonna use that song I sang, they're gonna have to strip it or at least shift it. I'm glad the principle is there for others to build on" (366).

However, liberty and equality are clearly insufficient in Reagon's view. She powerfully develops the ways in which (in the absence of receptive generosity) liberty and equality can be incorporated into strategies of assimilation and denial. Woman-identified women are equal so long as black women enter and remain within the bounds of the identities of lesbian white women "with a little color to it." But if black women express resistant differences, the equality of woman-identified women can easily proliferate a hierarchy between a privileged group and those with "false consciousness." Struggles in the name of liberty can easily acquire shades of assimilation, imperialism, or oblivion. Or, in the agonizing heat that builds around these tensions, equality and liberty might take a turn toward ethical and political indifference.

So there are reasons to doubt the political sufficiency of the ethical standpoint suggested by Laclau and Mouffe. Without a seductive account of the agonizing grandness of receptive generosity in the context of substantial difference, their position may lack the ethico-existential comportment necessary to sustain the political openness, tension, and ambiguity demanded by coalition politics within pluralist democracies.

One must wonder whether an openness which is to emerge simply through their juxtaposition of two logics of closure is sufficient to sustain admirable engagements, given Reagon's account of the dangerous, threatening, disruptive, and frightening character of coalition politics. When the immense pressures of coalition politics come to bear, do Laclau and Mouffe finally have a compelling ethical response when someone asks: "Why not seek simply to assimilate the other?" or "Why not seek to separate entirely?" (provided these options are strategically plausible in a given instance). There is little reason to be hopeful here. Laclau and Mouffe have no ethical account that would draw us toward and animate our engagements with these difficult others, no soliciting heights: their own project, in spite of itself, contains the seeds of other-assimilation and other-indifference.

In light of these dangers and difficulties, I think it is very significant that when Reagon explicitly reflects on the ethical direction that ought to guide our lives and "turn the century," it is *giving* which appears to be her highest virtue and which keeps drawing her to others *as others*: "But most of the things you do, if you do them right, are for people who live long after you are long forgotten. That will only happen if you give it away. Whatever it is that you know, give it away, and don't just give it on the horizontal. . . . [G]ive it away *that* way (up and down)" ("CP," 365). Without a *generosity* born in our efforts to receive the other *as other*, our gifts wither and equality and liberty will likely take up strategic positions within imperialist identities that assimilate, smother, or explicitly deny otherness. Generosity, as practiced in the efforts to grapple with core-threatening differences in coalition politics, just isn't sufficient to sustain one's life in coalition work in an uninterrupted manner: "You don't get fed a lot in a coalition. In a coalition you have to give, and it is different from your home. You can't stay there all the time. You go to coalition for a few hours and then you go back and take your bottle wherever it is, and then you go back and coalesce" (359).

You can't stay there (in the midst of the most agonistic difference). But generosity is a vital virtue that keeps one coming back for more—not only because in the absence of receptive generosity we sink into mindless mediocrity and subjection; but also because, when the grandness of giving and receiving occurs, "That's all you pay attention to: when that great day happen. You go wishing every day was like that" ("CP," 368). Every day isn't like that, but the experience and the wish call us toward future paths of giving and receiving. Thus called, we discover that the "move beyond Marx" requires for its desirability and success an agonistic move *toward* Marx.

Habermas, Adorno, and Radical Plural Democracy

I shall avoid discussing the foundational positions that separate Laclau and Mouffe, Habermas, and Adorno. Adorno, on my reading, is not as decisionistically contingent as Laclau and Mouffe, but he acknowledges far more contingency than Habermas. But what are the possibilities and pitfalls of Habermas's and Adorno's ethical stances in relation to a political terrain characterized by radically diverse social movements? Does either theorist offer an ethical vision that might help animate diverse social movements without inadvertently fomenting the dangers of imperialism, assimilation, narrow identity politics, or apoliticism? Can either offer a soliciting height that might draw diverse groups together in the form of many summits rather than in a single, unified peak?

Habermas's position begins to address some of the problems identified in relation to Laclau and Mouffe—even as it exacerbates others. Habermas enjoins a consensual movement toward others that opens onto self-transfiguration through learning with others that moves us significantly beyond the more assimilationist or imperialist resonances of the logic of combination in Laclau and Mouffe. Thus he writes: "The moral point of view calls for the extension and reversibility of interpretive perspectives so that alternative viewpoints and interest structures and differences in individual self-understandings and world-views are not effaced but are given full play in discourse" (*JA*, 58). Similarly, diverse social movements are "required" to bring the subjugated "voice of the other" into the political conversation (14–15). Like all communicative action, this intersubjectivity of the public spheres of different groups gives rise to a "virtual center." "Polycentric projections of the totality—which anticipate, outdo, and incorporate one another—generate competing centers. Even collective identities dance back and forth in the competing flux of interpretations . . . more . . . a fragile network than . . . a stable center of reflection" (*PDM*, 359). We have an image of reflective, edifying *coalescence* which contains an ideal of learning from others that in some respects exceeds imperialist or assimilative unification to a greater degree than Laclau and Mouffe's more stategically based "logic of equivalence."

Of course, Laclau and Mouffe seek to check the logic of equivalence. Furthermore, they recognize an aspect of violence and imposition, to which Habermas's rhetoric of "reaching" a consensus is insufficiently sensitive. Yet they come up short on an account of ethical unification that would both solicit our gathering for more than strategic reasons and inform it so that even *as unification* it is more cognizant of diversity and transfiguration. This insufficiency limits both the power and the desirability of their articulation of the unifying movement.

Furthermore, Habermas articulates an integral moment of social recognition associated with our efforts to withdraw from or resist extant social configurations (even if he errs in his singular emphasis on this moment). Thus he provides a vantage point from which the logic of autonomy and liberty can be understood as aiming toward more than the horizon of "auto-constitutivity," thereby checking the impulse toward withdrawal and indifference that Laclau and Mouffe's understanding of liberty tacitly overemphasizes. By infusing autonomy and liberty with an aspiration of "returning toward others" that animates even our turning away, Habermas articulates something that seems desirable given the stresses Reagon describes. Further, he construes the striving for idealized recognition as overflowing the bounds of any extant group identity. By theorizing all identity to be instituted differentially, Laclau and Mouffe are situated on an ontological terrain having profound ethical possibilities, but thus far they have largely left such possibilities undeveloped.

Hence, bracketing Habermas's shortcomings, one can discern a number of significant virtues with regard to the political problems Reagon raises. In other ways, though, communicative ethics falls short of the political challenges she identifies.

First of all, there is an insensitivity to the violence that accompanies thinking, especially thinking which is as singularly consensual as Habermasian moral discourse. I will not replay here the critique discussed above, but will focus instead on the political dimensions of this problem as related to the interactions among diverse social movements.

Discourse ethics is guided by a principle of universalization, the aim of which is to draw people toward "a general will that has absorbed into itself, *without repression*, the will of all" (*JA*, 13). With this principle in mind, Habermas embraces a deontological privileging of issues of "justice" over issues of the "good." The former issues "compel the participants to *transcend* the social and historical context of their partic- ular form of life and [their] particular community and adopt the per- spective of *all* those possibly affected" (24). The assumptions here are (a) that truly moral questions can be detached from all local identities and narratives by using "a knife that makes razor-sharp cuts" and (b) that they "admit of just one valid answer" (59).

The lack of any sense of tragedy, ambiguity, or humility in these passages is striking, as is the way Habermas avoids juxtaposing more agonistic sensibilities to encourage a moral dialogue more likely to contest, illuminate the damages of, slacken, transfigure, and limit the consensual imperative. From another angle, Steven Lukes has ques- tioned the relevance of Habermas's idea of rational consensus insofar as its achievement unrealistically requires homogeneous or abstract par- ticipants, or those who are readily willing to become such, and thus fails adequately to engage actual participants and their particular per- spectives and problems.[23] Similarly, Thomas McCarthy questions the sharpness of the edge between the just and the good, as well as the possibility of moral consensus under conditions where "there are fun- damental divergences in value orientations."[24] Yet, leaving aside a direct discussion of these questions of possibility, relevance, and a self-reflec- tive tragic sensibility that might transfigure one's yearnings—or, rather, approaching them from a different angle—I am interested in the politi- cal effects of this moral consensuality *as a singular demand.*

It seems likely that many groups, subjugated and marginalized his- torically and now beginning to develop power and a public presence, will be unmoved by Habermas's singular consensualism regarding moral issues and will perceive it as a form of subjugation to be resisted. True, communicative ethics makes a space for diverse voices in public moral discourse. Yet the activity those voices are called to perform in this space is pushed in the direction of "*transcend*[ing] the social and historical context of their particular form of life and [their] particular community [to] adopt the perspective of *all* those possibly affected." Concerning moral questions, then, diverse groups are welcomed into public spaces so that they might strive to abandon themselves. The consensus toward which they strive is to be "*nonrepressive*," but the *one- directional* pressures of striving are themselves likely to be perceived as repressive. The Habermasian position greatly underemphasizes any sense of the importance of our recalcitrant specificities for any moral engagement; the importance of the agonistic commingling of *different* narratives in the unending effort to illuminate, transfigure, and make sense of our lives and of "what we ought to do"; the need to cultivate, explore, and develop our *particular* modes of being in ways that may at once have moral significance *and* resist assimilation to an abstracting universalism, given singular a priori privileges; and, finally, any sense that the normative and the evaluative—the just and the good—may be so inextricably entwined as to infuse our consensual efforts with ambiguity.

These sensibilities are politically vital insofar as many groups and social movements are striving to explore and articulate differences that have been subjugated. They see these differences as sites of fruitful possibility—not just "evaluatively" but "normatively." A position that cannot resonate with this yearning is not likely to be politically effective or compelling. Indeed, it is likely to be resisted. Moreover, such a position transgresses the specificity and aspirations of social movements in an ethically problematic manner.

Faced with these resistances, as well as with the taxing tensions that Reagon describes, Habermas's position is likely to fall short in another sense as well. How will Habermasian selves—saturated with consensual strivings and demands for "one valid answer," insistent that communication be "straightforward and unambiguous"—fare in the significantly agonistic milieu of coalition politics? Would discourse ethics sustain political activity in such a milieu? Or does it rather tend to set us up for radical disillusionment and withdrawal? Might not an ethical opening that has a greater sense of the value of our unresolved engagements with those who are very different from us, a moral sensibility that embraces a significant degree of agonism, ambiguity, and distance as itself integral to moral activity, be better prepared for such difficult engagements?

Habermas would not be left speechless by these objections. In his recent book, *Justification and Application* (especially in the essay "Remarks on Discourse Ethics"), Habermas tries to limit the insistent character of normative consensuality and make a space for neo-Aristotelian reflections concerning those diverse forms of life which serve as a locus of evaluative identities and narratives. In response to my critique of the potential violence lodged in discourse ethics in relation to emergent groups, Habermas might point to his response to McCarthy, acknowledging that in the context of pluralization and multicultural societies it is "ever more improbable . . . that we will agree on shared interpretations in. . . . disputes" over needs, descriptions of ourselves and the world, and so forth (*JA*, 90). Thus, where what may have appeared to be a moral issue (e.g., abortion) proves to be "inextricably interwoven with individual self-descriptions of persons and groups" (59), it might transpire that the issue is in fact not a moral one at all— that is to say, not amenable to a resolution in terms of general interests. Where "it is possible to deduce from the inclusive outcome of the practical discourses" that the issues concerned are not morally resolvable, the moral question of generalizable interests then shifts to a "more general level" having to do with "how the integrity and the coexistence of ways of life and worldviews that generate different ethical conceptions . . . can be secured under conditions of equal rights" (60). Hence the realm of rationally resolvable, consensus-governed questions tends—in an empirically open manner—to reduce to a few issues like equal rights, democratic processes, legitimate conditions of compromise formation, and so forth. With this move toward reigning in and thinning out consensual demands (i.e., toward a greater space of "letting be"), worries about the ethically and politically untenable transgressions of consensualism might seem overwrought.

Yet Habermas's present position is unclear at best and is probably at odds with many of the ethical, ontological, and epistemological positions discussed in Chapter 3 above. His "clarified" position seems to drop the claim that the process of argumentation tames and enlists the world-disclosive (roughly equivalent to "evaluative" above) dimensions of language through context-transcending idealizations which develop "independent logics" and "learning processes" which "master problems." Now the world-disclosive functions seem to overpower and short-circuit logic, learning, and consensuality itself—except in terms of minimalist formal procedures and "letting be." The relation between the two dimensions is very unclear, however. We are left to "deduce" that evaluative discrepancies trump moral-consensual logics in the presence of "inconclusive outcomes." But what counts as an "inconclusive outcome"? How are we to tell when to reduce the thick consensual insistences and pressures lodged in Habermas's theory of communicative action, discourse ethics, and the self? And what happens when we "deduce inconclusive outcomes" concerning precisely such questions as an insistent consensual morality, the priority of justice over the good, and the relation between answers to these questions and the evaluative orientations of different forms of life? Do we not in *these* questions encounter ambiguities that are left mostly unacknowledged by Habermas? As it stands, we are left with a theory of thickly and deeply articulated consensual pressures that are ethically and politically problematic, topped with an ad hoc reigning-in of these basic demands.

This unaddressed tension is located not simply between earlier texts and "Remarks on Discourse Ethics" but within the latter work as well. Rather than illuminating and perhaps transfiguring the undersides of different concerns through tensional juxtaposition, these inconsistencies appear to *conceal* undersides so that the dominant motifs of the communicative framework can continue full steam ahead. Under the cover of liberal limits, Habermas continues to argue for the sharp separability of questions of justice from those of the good as well as the priority of justice over the good; and he can continue to deploy the insistent and singular rhetoric of compelling participants to transcend particular forms of life, proclaiming that moral questions have "just one valid answer." Thus he maintains the highly pressurized currents of his normative position while his text covers the questions that might arise with soothing reassurances (but from where?) that truly moral questions are few, that inconclusivity short-circuits (or reduces to tolerance) the compulsions of context-transcending participation, and that evaluative questions may lie at the bottom of most questions that have been misconstrued as moral.

Politically, then, Habermas's deepest currents remain unchanged and probably continue to lack much motivational power for many new social movements and groups. However, if we consider separately this newly accented voice for "letting be" diverse groups, significant political questions remain. Surely the impulse simply to "let be" has much to recommend it and is preferable to many of the ways in which groups relate to each other. Yet my suspicion is that "letting be" is a moral outlook which is nurtured precisely *through engagements* that involve *more than* the relative indifference of parallel, mutually uninvolved co-existence: namely, receptive engagements where one recognizes at a distance something profound, even though it is not one's own; engagements through which one is powerfully transfigured, though probably not in the sense of self-transcending convergence or mutual mergence. In other words, "letting be" is an active virtue obliquely cultivated. My guess is that these partly cooperative, partly agonistic relations form a significant aspect of those practices which might solicit the virtue of

"letting be" *as part of* an ethical constellation. Insistent consensual striving probably does not; backing toward a simple acceptance of auto-
constitutivity probably does not. Furthermore, it is questionable whether
"letting be," as Habermas articulates it, could do much to sustain and
animate the difficult engagements and tensions of the coalition politics
described by Reagon. As with Laclau and Mouffe's liberty, a simplistic
"letting be" might be more conducive to a withdrawal from politics.

In light of the shortcomings of Laclau–Mouffe and Habermas vis-
à-vis the difficult terrain of coalition politics, reading Adorno might
contribute something. Adorno, however, is widely regarded as one who
has lost hope in political activity—a claim which is largely true.
Adorno is also said to ontologize hopelessness in his critical theory—a
response to Hitler, Stalin, and what he took to be the near "totally
administrated society" of "late capitalism," based on a fusion of corporate capitalist power, a hierarchical administrative state, and a culture
industry that has spread subjugation deeply and extensively. How can
he be of help politically?

A lot of contemporary analysis compellingly illustrates—in part
by illuminating the contradictions, tensions, and ambiguities in our
world—important possibilities for progressive transformations that are
absent in Adorno's practical assessments. Yet it is historically an open
question whether or not Adorno's general analysis was too pessimistic
or whether it pretty much got things right. At any rate, my reading
resists the idea that Adorno's theoretical work necessarily issues in this
pessimism. Rather, I interpret his work as an ethical constellation for
engaging the world. Insofar as the spaces, possibilities, and conditions
for this practice hinge on the character and extent of our political actions, then so long as political transformation is historically an open
question[25] the "morality of thinking" must strive to articulate itself in
directly political ways. I go beyond Adorno here, but my reflections are
in keeping with the comment in the 1969 "Preface" to the *Dialectic of
Enlightenment* that "today [the question] is more [one] of preserving
freedom, and extending and developing it" (x). For reasons having
partly to do with the demise of the politics of collective subjectivity (as
well as for other reasons, which I shall discuss in relation to Habermas's
analysis of contemporary capitalist democracies), this question must
significantly address the issues and difficulties of social movements and
coalition politics.

Adorno's work points beyond the political pitfalls of Laclau and

Mouffe and Habermas. Adorno provides us with an image of coming together that exceeds imperialism, assimilation, or the logic of equivalence. Unlike Habermas, though, Adorno does not construe learning solely in terms of consensual convergence. Rather, he envisions endless transfiguring relations of receptive generosity: "communication of what [is] distinguished" and "distinctness without domination, with the distinct participating in each other."[26] Such participation is induced in part by a pacific consensuality. Yet Adorno's consensuality accents dimensions of distinctness and distance ("distant proximity"). A generous receptivity draws us beyond our individual and group forms of life to engage others, to mark and be marked by others, to transfigure and be transfigured. However, this movement is grasped not as a convergence toward a single identity, but rather as an open-ended articulation of freedom and well-being with inexhaustibly distinct and nonidentical others. Being thus drawn outward opens up the possibility of coalition politics.

Simultaneously, Adorno juxtaposes with this outward-moving sensibility a strong sense that resistance to the world's pressurized immediacy also demands a concerted effort to think *in the light of* our specificities; that it demands efforts to think through the historical particularities in order to illuminate possibilities for understanding, freedom, and well-being that are eclipsed by an overly "administrated" and normalized social world. Though Adorno never wrote explicitly about the possibilities of group differences, clearly efforts to consider this question from the perspective of negative dialectics must recall the importance of immanence as a moment in his thinking. He draws us to "the question [of] how a thinking obliged to relinquish tradition might preserve and transform tradition. For this and nothing else is the mental experience" (*ND*, 54–55). This "mental experience" is exemplified by Adorno's relation to the tradition of thinking in which he was most steeped: namely, German idealism.[27] Adorno is powerfully aware of the constraints, compulsions, and blindnesses of this tradition. Yet, recognizing that "knowledge as such . . . takes part in tradition as unconscious remembrance [and that] there is no question which we might simply ask, without knowing of past things that are preserved in the question and spur it" (54), he strives for an opening beyond the limits of idealism *through* the latent possibilities of its horizons. He seeks to "break the compulsion" within this tradition by developing the "energy and insights" contained in such compulsion.

Thus, Adorno's constellation solicits a cultivation of tradition, not as a frozen impermeable identity but as a dynamic site of possibility in spite of the subjugative dimensions of any history. In addition to being epistemologically and ethically compelling, this approach creates a greater possibility for engaging diverse social groups than one finds in Habermas. Such a perspective resonates with (and solicits) the aspirations of many groups to cultivate the possibilities of their distinct traditions. Nevertheless, it does so in ways that encourage the opening of possibilities for articulating receptive generosity beyond group borders. And, as we've seen, Adorno *also* solicits an interrogative "stepping out" in tension with the moment of immanence.[28] Doubtless this ethical stance will be strongly resisted by those engaged in a narrow, essentialist, or wholly self-serving identity politics. Yet much of the politics involving questions of identity, group difference, and so forth is not definitively "closed" in these senses, and thus it is plausible that Adorno's morality of thinking might pull diverse groups toward a coalition politics of receptive generosity.

As Adorno calls us together, he does not lead us to expect that the ensuing relations will be easy and harmonious; rather, in a manner well-suited to the demands of coalition politics, he instills in us a sense of the mutual resistance beings tend to present to each other, a sense of the blindness and damage of our relations, of the extent to which "discord is the truth about harmony." Moreover, he cultivates a sense of the *value* of agonism itself, through which we might gain a sense of our limits and move beyond reified closures to elaborate freer and richer possibilities. Of course, even though Adorno reminds us that agonisms are often entwined with subjugative *antagonisms* which we must reject, he nevertheless locates the practices of freedom, well-being, and generosity not primarily *beyond* resistance and tension but precisely *in* such relations. He provides us with a sensibility more capable of embracing the difficulties of coalition politics.

Reading Adorno in relation to the demands of coalition politics, I am suggesting that his embrace of immanent development, his solicitation of an interrogative movement toward others which is less narrowly consensual than Habermas's, and his esteem for agonism as essential to the virtues of ethical activity all enable his position to *engage* the participants, to *help sustain* our political activity, and to *transfigure* it toward the development of a more admirable politics of receptive generosity. My reading suggests an ethical constellation which is more conducive than discourse ethics to the animation of a vibrant civil society. Adorno weaves a web of insights, solicitations, and questions that cut across us from myriad angles and that disrupt simplifying and damaging identities. He calls us to a profound appreciation of the pregnant specificity of ourselves and all beings.

Or, he does this *if* we read him as I have been doing. In truth, though, his texts have very uneven effects, only occasionally provoking the responses and activity—even at the level of theory—that I am suggesting. Ultimately I am unwilling to attribute this discrepancy wholly to a failure in his readers. In fact, a good many subtle and insightful people, many of whom are not too distant from my philosophical, ethical, and political orientations, will undoubtedly remain skeptical that my reading corresponds to the Adorno they know. I am convinced (though I am my easiest victim) that I have traced important strands in Adorno's intricately woven textual web. But to me this question of "accuracy" seems, finally, unimportant. More important is the philosophical and political value of the constellation. If it illuminates our questions in a compelling manner, perhaps negative dialectics must be rewritten. What I have in mind is not a new text that would replace the old, but rewriting(s) that might join it, creating tensions and friction that could light up desirable possibilities and illuminate dangers less visible in the shimmering starlight of Adorno's melancholy sky. Such a project would weave a constellation into the becoming-and-growing constellation of constellations which negative dialectics must be—if it is to *be*.

This book is an effort at such a rewriting (so runs my confession after the deed). And if I have not written much since my chapter on the Kant of agonistic giving and receiving in the Third Critique, that is because he evokes the spirit of, and is already present throughout, my rewriting of Adorno's work.[29] *This* Kant energizes my interpretation, which aims to alter some of Adorno's directions and to accent some of the buoyancy and sense of vibrant possibility one finds within Kant's discussion of genius and aesthetic ideas. In my imagination, this Kant and this Adorno stand in proximity, sharing themes of nonidentity, violence, tragedy, the movement of thought beyond itself, dialogue, language on the edge of identity and nonidentity, giving and receiving, agonism, and so on. Despite striking similarities, however, theirs is nevertheless a *distant* proximity. The distance concerns "spirit," which Kant (let us recall) describes as the "animating principle in the mind."

Put simply, this spirit of Kant's has a certain buoyancy, a levity about it, that one finds only rarely in Adorno (and even then floating with a melancholy eye to the "wounded coast"). The material that this buoyant spirit employs to "animate the soul" is proximate in thematic form but more distant in terms of the density and weight of its flesh. And this distance is discerned in the rhetorical figures that lighten and raise the spirit of Kant's text: "purposive momentum," "prompts the imagination to spread," "quicken," "rapidly passing play," and so forth. Though we do not hear of "wings" until Kant writes of taste "severely clipping" them, Kant's thinking here is indeed winged.

Perhaps I have come to my rewriting of Adorno with wings; and not so much that he might fly a bit more, for Adorno's thinking surely flies. Rather, wings to add a bit more buoyancy to his flight in a effort to increase its power to lift other flights. Buoyancy has its dangers, as this student of twelve-tone and atonal composition surely warned. But so does melancholy. Even "after Auschwitz." The two should be linked, share their affinity, criticize each other.

Buoyant or not, one thing seems certain: if negative dialectics cannot discover languages capable of engaging more people than it has, its ability to animate social movements and civil society will remain slight. For some time now, there has been a growing resentment against academic discourse; I have no wish to join those ranks. Philosophy requires spaces for modes of articulation which are not easily comprehensible to those less engaged in such reflections (though perhaps Adorno erred on the side of too little concern for communicability). The problem begins when those engaged in such spaces fail to construct—and even choose to sever—bridges of communication between themselves and a larger public. This is our present unfortunate condition, in my view. Certainly the destruction of such bridges comes from a multitude of sites unrelated to political theory, and certainly there are high risks involved in all such crossings. But the risk of staying put is far greater. There is no generosity, no receptivity, until these questions become a central concern of critical theory. Adorno knew this (but is of little constructive help here). My rewriting in the present book makes no contribution to this task.

In our move away from Marx to Laclau and Mouffe, Habermas, and Adorno, an important theme from Marx returns, though it is differently construed. For what is lacking in Laclau's and in Mouffe's radical liberalism and discourse ethics, with respect to the challenges and possibilities of coalition politics, is a compelling response to the questions of receptive generosity that animated Marx. In this void we are likely to remain entangled in strategies of imperialism, assimilation, or indifference.

Of course Marx's evocation of a harmonious giving-and-receiving without recalcitrance, through which human beings are to produce and appropriate all of being as their home, courts disaster beyond anything likely to arise from discourse ethics or radical liberalism. We must therefore concern ourselves with the possibilities of receptive generosity on a terrain without collective subjectivity and the resistance-free giving and receiving with which it was to be entwined. We must bestow on receptive generosity an aura of inexhaustible questionability. To dream that society will one day be able to "inscribe on its banner" the fact that receptive generosity no longer exceeds the world as a soliciting agonistic question is to dream of the death of giving; and giving and receiving begin to die with this dream.

Adorno lucidly ponders this death, and he attempts to reconstitute life as possibility—not by forgetting what was profound about the dream, but by radically transfiguring it in the context of his ethical constellation. Yet his constellation of solicitations toward an agonistic receptive generosity is largely removed from the terrain of politics. I have argued that negative dialectics has significant potential to reengage, energize, and transfigure a coalition politics of movements and groups diversely drawn together by an ethic of receptive generosity. Let us reopen the question of political possibility.

Preliminary Thoughts on Civil Society and Political Economy

Questions of receptive generosity were central to Marx's inquiries into political economy. After Marx it is only with a studied myopia that the ethical imagination can ignore political economy. For, whatever the substantive weaknesses of some of Marx's own analyses, there can be no doubt that political economy is a crucial location of problems and possibilities that require transformation if ethical practice is to advance in contemporary society. Influences of the capitalist market on the development of culture, self, community, knowledge, morality, power, and ecology push us in directions which are at odds with anything remotely resembling the orientations of the present work.

Recognizing the impossibility of addressing these issues here, I do want to consider a way in which social movements might *begin* to engage, transform, and move beyond contemporary state and market systems in light of Adorno's ethical concerns. I shall begin with an overview of Habermas's understanding of our political possibilities in late-twentieth-century capitalist societies. I then proceed to think with and against him.

Habermas cautions against simplistic moves from discourse ethics to a full-blown substantive morality and politics. Yet, from his *Structural Transformation of the Public Sphere* through the recent *Between Facts and Norms: Contributions to a Discourse Theory of Law and Democracy*, Habermas's ethical and political philosophies have been deeply entwined. With his view that political theory is much more entangled in historical contingency than are transcendental pragmatics and discourse ethics, he warns us against simplistic translations and overly transcendental claims in the realm of politics. But he does not mean to dissuade efforts to consider politics in light of discourse ethics. Here I sketch Habermas's thoughts over the past dozen years on politics and economics, during which he has moved a long way from the more radically democratic socialist totality which vaguely formed the normative political background of his writing during the 1960s and 1970s.[30] It is perhaps also possible to see a recent waning of the radically democratic moments in his writings. Nevertheless, it is with the great strands of continuity that I am chiefly concerned.

Habermas claims that his move away from the Marxian ideal of a radically democratic, totally self-administrating political economy stemmed from an increasing awareness of the immense complexity of modern, functionally differentiated societies. Such complexity requires systems that function by means of steering media like money and bureaucratic power in order to deburden the communicatively structured lifeworld, which is simply unable on its own to address and coordinate modern activity. He argues that state bureaucracies and economic systems "can no longer be transformed democratically from within . . . without damaging their proper systemic logic and therewith their ability to function."[31] We need these systems. So, instead of seeking to displace the state with some form of radically democratic governance which would then thoroughly subject economic life to collective power,

Habermas now suggests a "separation of powers" in which state, economy, and a civil society of diverse public spheres are each granted a degree of automony. The new "radical democratic" goal within this context is to "erect a democratic dam" against state and economic colonization of the lifeworld and to make these systems more responsive to the communicatively generated guidance of "the practically oriented demands of the lifeworld."[32]

This conception of an offensive and defensive assertion of democratic power in the context of powerful administrative and economic systems hinges on the enhancement of communicative practices at two related levels. First, discursive democracy takes the form of a constitutionally governed set of procedures and rights that formally guarantee the possibility of a state that legislatively operates (internally) according to, and is responsive (externally) to, the "communicative force of production." Second, in order to have substance, discursive democracy requires a vibrant civil society with multiplicitous sites that nurture public discourse. Let's turn to each of these levels and then analyze their relation to the systems governed by steering media.

Habermas argues that democratic constitutional states and laws arise largely in response to problems of social integration which result from the disenchantment and pluralization of lifeworlds as well as from the proliferation of large areas of strategic action in economic systems. As traditional lifeworld and institutional sources of authority wane, the tasks of social integration shift to the shared agreements that can be reached communicatively by members of a society. Yet as the unquestioned background assumptions of the lifeworld grow thinner and less extensive, the discursive resolution of problems becomes more difficult and the risks of dissent become greater. In this context of overburdened communicative mechanisms of social integration, a level of uncertainty and chaos can emerge and undermine norm-governed behavior. However, modern law acts here to constrain this destabilizing chaos in both a "normative" and a "positive" sense. In terms of what Habermas calls "facticity," modern law functions (through the imposition of state-guaranteed enforcement) to stabilize actual behavior and expectations. Law constitutes regularity by factually removing (through enforced observance) both the necessity and the possibility of having to reach a new agreement at each instant. This positive stabilization, however, is entwined with a normative validity: "Law borrows its binding force . . . from the alliance that the facticity of law forms with the claim to legitimacy" (*FN*, 38–39). This latter claim is rooted in the law's origin in, and further generation of, institutions which instantiate communicatively rational self-legislation by means of a constitutionally differentiated democratic state—a state which is designed to thwart the penetration of power relations generated elsewhere in society. The intertwining of facticity and validity, Habermas argues, can be seen in the way that the factual presence of law allows norm-governed behavior to manifest itself regularly in modern societies while, in turn, this positivity draws out its own stabilizing legitimacy in connection with procedures approximating communicative rationality.

Habermas seems more or less content with constitutional forms like the one bestowed on the "fortunate heirs of the Founding Fathers" (*FN*, 2). Yet this backward-glancing reference to a "foundational act" which marks a beginning we continually return to in interpretation, is entwined with a forward-looking view: "the horizon of expectation opening on an ever-present future" (*FN*, 384). For Habermas, "the constitution [is] an unfinished project . . . a delicate and sensitive—above all fallible and revisable—undertaking whose purpose is to realize the system of rights *anew* in changing circumstances, i.e., to interpret the system of rights better, to institutionalize it more appropriately, to draw out its contents more radically" (384).

Habermas distinguishes his own position both from the liberal view of democracy, characterized by a "compromise of interests" and balance of power among prepolitical individuals atomistically endowed with universal rights, and from the republican view of Rousseau in which democracy is the political self-determination of society's collective will. Discursive democracy is centered on an unobstructed process of deliberation and decision-making based on "rules of discourse and forms of argumentation that borrow their normative content from the validity basis of action oriented to reaching understanding" (*FN*, 296–97). Discursive democracy understands the constitutional state as "a consistent answer to the question of how the demanding communicative forms of democratic opinion- and will-formation can be institutionalized" (298).

Yet, as noted, while the deliberative processes that occur within the parliamentary institutions of the democratic state are necessary aspects of discursive democracy, they are insufficient and depend on extensive networks of informal public spheres in civil society. With John Dewey, Habermas evaluates democratic politics as a "two-tiered" phenomenon that can be considered to function well only when the informal level is thriving and is programming responsive formal institutions.[33] While it is the formal institutions alone which are to gather and distill societywide communicative streams into pools of legislative action that guide the administrative systems, the informal public spheres are nevertheless vital to the overall deliberative process. They provide a "context of discovery" which, free of procedural regulations, is far more "wild" and "unrestricted," ranging "more widely and expressively" (*FN*, 307–8). Because legislative bodies operate according to formal procedures and under time constraints, they are poorly suited to identify problems that are latent and not yet articulated. Far more sensitive are the informal public spheres, where questions and struggles concerning need, recognition, identity, etc. can be pursued over broader and more uncertain terrain. Gradually, as issues are articulated and expressively dramatized, they begin to infiltrate parliamentary discourse and decisionmaking. "It is usually a long road until the controversial contributions on such issues . . . adequately articulate the needs of those affected. Only after a public 'struggle for recognition' can the contested interest positions be taken up by the responsible political authorities, put on the parliamentary agenda, discussed, and, if need be, worked into legislative proposals and binding decisions" (314).

The discussion thus far concerns an idealized, normatively legitimate order of things. Yet Habermas is aware that the actual picture in contemporary capitalist democracies is often quite different. Frequently the economic system colonizes the lifeworld and utilizes massive resources to systematically distort public spheres and political power in contradiction with the ideals of democracy. Similarly, the administration systems of the state often colonize the lifeworld and develop modes of operation that are recalcitrant to democratic will-formation. Furthermore, the "culture of privatism" fostered by contemporary capitalism is at odds with the ideal of a flourishing public life. Habermas responds to these dysfunctions in two ways. First, he claims that the normatively legitimate order he describes is—to varying degrees and despite significant distortions—an important dimension of modern democracies. Second, he reconceptualizes the ideals of democracy, away from the idea of a transparent state and toward the idea of "boundary conflicts."

While Habermas recognizes that, under normal contemporary circumstances, informal public spheres are distorted and occluded by political-economic power which establishes predominantly one-directional communication flows, he argues that in times of crisis these flows can and often do change direction in ways that allow the informal realms of civil society to become increasingly influential. Citing the extent to which such issues as the arms race, ecological crisis, poverty, and feminism have been raised by individuals, associations, and social movements in ways that have ultimately had a substantial impact on national political discussion and policy, Habermas claims that (though often docile) the public sphere in contemporary democratic-capitalist societies is capable of grassroots mobilizations that instantiate efficacious—if discontinuous—forms of discursive democracy.[34]

Yet this grassroots democratic practice is in most "normal situations" eclipsed, or at least compromised and co-opted, by political and economic systems that colonize the lifeworld. While it is possible for public spheres to influence the democratic state (which in turn may influence and regulate the economic system), in modern complex societies Habermas does not understand this influence as operating even normatively through the paradigm of a transparent, democratically self-organizing totality.[35] This is so not simply because of the "pluralization of the lifeworld" but also because systems tend both to accrue power which is recalcitrant to external imperatives and to view the "outside" as a resource environment for the system's own ends. Unless we are prepared to do without systems (and for Habermas this is inconceivable), we must understand democratic influence according to "the model of boundary conflicts—which are held in check by the lifeworld—between the lifeworld and two subsystems that are superior to it in complexity and can be influenced only indirectly, but on whose performances it at the same time depends" (*PDM*, 365). The aim of public spheres in this context should be the "building up [of] restraining barriers for the exchanges between system and lifeworld and . . . building in sensors for the exchanges between lifeworld and system" (365). This task involves diverse social movements and must combine the power of growing solidarity with "intelligent self-restraint." For even though systems must be restrained and made more sensitive to the needs and desires of the lifeworld, in Habermas's estimation it is crucial to "leave intact the modes of operation internal to functional systems and other highly organized spheres of action" (*FN*, 372). While he has not developed this notion in detail, clearly Habermas is concerned that radical democratic movements with an insufficient grasp of the vital role of systems in societal steering might "kill the goose that lays the golden egg" of stable democracy and economic productivity and coordination. The aim of radical democratic movements must be the maintenance, separation, and rebalancing of the powers of the political system, of the economic system, and of the lifeworld solidarities—with a decided shift in favor of the latter.

Adorno opens up possibilities in these three domains, particularly with respect to the relations between social movements and systems.[36] I start, then, with a brief consideration of relations within civil society as they might develop in the light of an ethic of receptive generosity.[37]

Habermas correctly notes that the vast network of informal discourse is wilder and frequently more sensitive than deliberation within procedurally rigid and time-bound institutions. He is also correct that informal public spheres have far greater capacities to discover unacknowledged suffering and to elaborate alternative possibilities regarding many issues. Yet, from Adorno's perspective, the effect of the assumptions and insistent consensuality of discourse ethics is precisely to *tame* the agonistic play of more world-disclosive discourses, thereby engendering the concealment of nonidentity and perpetuating violence toward subjugated modes of being. In contrast, Adorno's ethical constellation—soliciting our awareness of nonidentity, tragic finitude, violence, and specificity; soliciting, too, our receptive generosity—is likely to cultivate and intensify that sensitivity to and that respect for otherness which is a potential virtue of all informal public spheres (as noted also by Emerson, Whitman, Dewey, and others in the democratic tradition of the United States).[38]

It is not just the vibrant character of the public spheres that might significantly hinge on a post-secular *caritas*, but perhaps the very *existence* of those spheres. Radical democrats, at least since Dewey's confrontation with Lippmann's democratic "realism," have had to contend with the striking *lack* of participatory political dialogue and activism in democratic capitalist countries. Habermas concedes that during extensive "normal" periods participatory public discourse tends to be elite-dominated and inactive, yet he offers examples of grassroots movements which illuminate the possibility that things could be different. It would be naive to suggest, given the long list of contemporary circumstances and power structures which suppress the development of active public spheres (as Dewey enumerates in *The Public and Its Problems*), that an ethical transfiguration alone could bring public spaces and movements to life; or that any ethical transfiguration could spread far and wide without a simultaneous transformation of many dominant power structures and practices. Nevertheless, ethical transfigurations do play an important role, and no movement to transform contemporary practice will do anything that does not bear the mark of the ethical frameworks of those involved.

The question, then, concerns what ethical solicitations might further stimulate active public spheres and movements to advance social conditions that are conducive to meaningful participation in political life? I think that both Habermas and Adorno have much to contribute here, but that the latter has some decided advantages over the former, two of which I discuss.

First, Adorno's ethical constellation solicits distinct and active voices in ways that significantly disrupt passive conformity. Negative dialectics' evocation of the need for "lingering eyes," which seek to see beyond the compulsions and pressures of codified perceptions and identifications, and Adorno's sense of the importance of thinking in *resistance* and *dissent* both incite us to active, awakening engagements with the world and with others. These moments however, so that they don't become another form of self-identical indifference, are juxtaposed with consensual solicitations to draw together our differences. Even though Habermas's consensualism certainly makes space for dissent, and even provokes it insofar as extant affairs fall short of the ideal, the pressures guiding epistemology, morality, and self-development nevertheless may constitute a sleepiness or, worse, a rather stifling kind of interaction which is far from conducive to vibrant and sustained engagement. A singular consensualism may resonate with and reinforce conformity in unexpected ways.

Second, Adorno, as I have rewritten him, significantly solicits the practice and movement of receptive generosity and generous receptivity as good in themselves, as a height and depth to which we are called. In a sense, this articulation of life is a kind of meta-good that Adorno thinks might infuse and inform other goods. By thus soliciting us, Adorno nudges us toward becoming more political selves; selves engaged politically *not only* to realize other ends but, especially in the context of diverse groups and movements, in order to open up relationships where we sense and participate in "the inextinguishable color that comes from nonbeing." Reagon and Adorno meet here. Might not such an ethic contribute to multiplicitous and more energetic public spheres, less prone both to dissolution whenever a sense of crisis passes and to elite and systemic management? Of course, participatory politics (especially coalition politics) is not just "that great day" which Reagon spoke about. As she tells us, such a politics is demanding, feels life-threatening, agonizing, frustrating—a place where one cannot stay. But it might well be a place more frequently and more powerfully inhabited.

Along with the possibility of a more animated "civil society" come questions concerning its relations with, and efforts to transform, the bureaucratic state and a market system where vast power is concentrated in the hands of gigantic capital. As we have seen, Habermas envisions these relations in terms of boundary conflicts wherein social movements seek to restrain systems from colonizing the lifeworld and to make them more sensitive to the demands of the lifeworld. His more recent work increasingly emphasizes self-restraint on the part of social movements so that they "leave intact the modes of operation intrinsic to functional systems."

Any effort to address these issues will require vast research and a patient, multifaceted experimental process. Nothing less is likely to give desirable form to radical aspirations. However, my rewriting of Adorno suggests a vision of political and economic possibility that varies significantly from Habermas's paradigm and warrants further inquiry. A brief discussion of some points of agreement provides a context in which to clarify a few points of divergence.

Habermas's case that there is a need for bureaucratic and market systems which function according to steering media is compelling.[39] I don't see how, short of a *massive* decentralization and dedifferentiation of nearly all realms of social life (which seems nearly impossible as well as undesirable), human beings could bear the transaction costs of production, coordination, allocation, regulation, program implementation, etc., let alone address all related issues, solely by means of communicative deliberation each step of the way. Furthermore, some degree of system differentiation between state and market steering also appears desirable in light of the massive bureaucracy that otherwise seems necessary in order to govern economic coordination and allocation, not to mention the extent to which bureaucracies quickly become resistant to external democratic will-formation. And precisely because of the turbulence and aggravation that Reagon describes, some version of (amendable) constitutionally organized checks and balances, formal procedures, and so forth is clearly necessary in central spaces of deliberation and legal-policy formation, in order to prevent the abuses of power to which humans are so prone.

Yet these points of general concordance provide a background for substantial contestation. One crucial divergence concerns the *extent* of Habermas's respect and restraint concerning systems: the extent to which he increasingly appears to want to "leave intact [their] modes of operation"; the extent to which the state still remains the central focus of his normative analysis of institutional democracy.

Similar questions have been raised by some who are closer to Habermas philosophically. Thomas McCarthy, for example, is concerned that "Habermas has taken over so much of the conceptual arsenal of systems theory that he risks not being able to formulate an answer to the question [i.e., what *forms* of representative democracy and public administration?] compatible with his professed political ideals."[40] By going too far in the direction of systems autonomy, writes McCarthy, Habermas risks radically truncating political participation. Calling on Habermas to consider the importance of noninstrumental grounds in our deliberations on these issues, McCarthy worries that because Habermas has taken "a more strongly theoretical direction" since the 1960s, he has bowed too deeply to the apparent—though illusory—theoretical power of systems theory.[41] McCarthy calls his mentor away from a general *theoretical* privileging of systems autonomy and back toward a more open-ended and empirical experimental approach, drawn by the radical participatory ideal he finds in the theory of communicative action.

However, I suggest that the problem (the rather closed and almost a priori feel to Habermas's privileging of systems) has deeper roots in Habermas's paradigm, and that understanding the problem in this way allows one to articulate a more intransigent vision of (and a horizon of concern for) noninstrumentality and radical democracy than is available from the theory of communicative action.

In Chapter 3, I showed the extent to which pressures of coordination are central to Habermas's understanding of the genesis and character of communication, morality, and the self: in short, the extent to which Habermas understands humans most fundamentally as coordinating beings. Though from early on Habermas has favorably juxtaposed communicative modes with instrumental-strategic modes of action, his understanding of the former has itself been so heavily overdetermined by instrumental concerns that one should not be surprised that instrumentality is now gaining substantial autonomy from communication. As McCarthy is aware, a very significant aspect of Habermas's debate with Niklas Luhmann in *Legitimation Crisis* centered on the question of whether democratization or nonparticipatory administrative systems could better address and avoid compounding issues of complexity in modern societies. In that text, still significantly under the sway of a radically participatory tradition, Habermas argued that empirically there were advantages and disadvantages to each mode of collective action, and he clearly leaned in the direction of participatory democratization. Yet his argument has always focused largely on the coordinating advantages that communicative action can secure in terms of learning, motivation, and legitimation. More recently, in *Between Facts and Norms*, the argument for the continued importance of communicative action is developed in terms of intersystemic coordination advantages made possible by the fact that systems remain anchored in a lifeworld which can communicatively mediate between them.

No transfiguration of ethical or political visions will eliminate these important questions. Nevertheless, the overwhelming emphasis on this kind of questioning and the extent to which we orient our political and economic directions in light of the outcome to these questions *is* highly contestable. My suspicion is that the Habermas of the late 1960s and early 1970s continued to be animated by a neo-Marxian background that, as we have seen, esteemed radical democracy for reasons which included but exceeded its coordinative capacities: reasons which concerned the historical incarnation of relations of receptive generosity, however vague, partly as a good in itself. Yet during this period, two related developments began to take place. The ethical idea of a beautiful receptive generosity withered along with the collective subjectivity that was its condition of possibility, and Habermas's ethical ideal was transfigured into the notion of solidarity articulated through communicative action. Simultaneously, Habermas began his monumental effort to clarify and articulate theoretically the communicative impulses that had guided his work from *Structural Transformation* forward. Increasingly he articulated a theory that, from Adorno's perspective, articulated "the spirit of the age," "what is called communication today," "the adaptation to useful aims." Issues of instrumental coordination gained centrality in Habermas's thinking. As they did, Habermas became particularly prone to "the seductions of systems theory"; for the latter provides, *in theory*, a level of squeaky-clean coordination of complex interactions whose communicative attainment is unimaginable. Henceforth, the role of communication is to guide systems in ways that rationalize their intersystem interactions and secure legitimacy and motivation.

McCarthy and others correctly identify the empirical weaknesses of systems theory and emphasize the potential learning capacities, intelligence, and coordinative virtues of political participation;[42] and such work is important to widening the possibilities for a politics accenting the practices and aims of receptive generosity. Receptive generosity, however, ought to have a solicitive power in its own right. If coordinated interaction is not *the* defining characteristic and *the* telos of human beings, then we can imagine participatory political interventions and disruptions of established systems that advance beyond exclusively coordinative concerns and limits. Our esteem for the steering capacities of systems is not reduced to zero from this perspective, for we *are* beings for whom coordination issues are of great concern. Nevertheless, if we are solicited to depths and heights that far exceed these concerns—as we are by the ethics of post-secular *caritas*—then these concerns can be politically contested, disrupted, dislodged from their occupation of the center of the discourses and practices that mobilize and constrain our energies.

If the dialogical practice of receptive generosity becomes more salient in our lives, and if we seek to orient the *outcomes* of substantial areas of collective action according to the sensibilities and aims entwined with this practice, then we expand the realm in which it might be desirable to diversely reconstitute, supplement, supplant, and invade systems in order to create spaces for less-restricted dialogues in the processes of acting—even where doing so may decrease coordinative capacities to some extent. We can begin to imagine supplementing and contesting the state with formal and informal strengthened public spheres, communities, and powers at the local, regional, and global level. We can imagine reconstituting economic power through a politics that invades the privacy and prerogatives of corporate boardrooms—diversely constraining, juxtaposing, supplementing, and, especially, *supplanting* the hegemony of capital with a variety of strategies for popular participation and more-generous outcomes.

As I have stressed, this wilder vision and solicitation calls for a "wild patience" (borrowing from Adrienne Rich): careful explorations and experimentation. I am *not* suggesting that we supplant Habermas's concerns for political and economic coordination. A politics of post-secular *caritas* harbors those concerns even as it exceeds them. Receptive generosity is empty without all sorts of coordination, careful calculation, efficiency, and so forth. Furthermore, any movement to transform our political economy will fall by the wayside if it does not address the vast array of important bread-and-butter issues. Insofar as mounting evidence suggests that learning, sensitivity, coordination, and efficiency are increased through participatory transformations, then it may frequently turn out that coordination and *caritas* suggest similar or overlapping political economic directions. Indeed, evidence from Robert Putnam's investigations indicates that intelligent and well-coordinated governance is compatible with highly contestory and conflictual political life: "*None* of these investigations . . . offered the slightest sustenance for the theory that social and political strife is incompatible with good government."[43] Still, it is likely not always to be the case that increased participation and coordination capacities will converge. At any rate, discrepancies and tensions will arise owing to the wilder nature of the dialogues solicited by a politics of receptive generosity relative to a communication singularly oriented by coordination.

Hence, I think we are left with a permanent and desirable agonism of concerns at the heart of political economy: a contestation over the extent to which coordination ought to be more central (and the structure, pressures, and character of communication and morality more as Habermas describes them) and the extent to which the solicitations of a post-secular *caritas* ought to be given greater sway (and the textures and aims of our engagements significantly different). If an ethics and politics of receptive generosity contains and contests coordination, it is also contested *by* it. This contestation must be carefully nurtured as a vital power within a more desirable political economy.

Post-secular *caritas* seeks paths *beyond* our political economy of endless growth and concentrated power that is waging war on people and on our planet. Here, there are pressing questions concern the nature and possibility of movements toward a political economy that communicatively "rationalizes" aspects of life (via more narrowly circumscribed discursive procedure and practice, more sensitized systems, etc.) in realms where we discern that coordination questions are utterly dominant. However, such movements must also make way for a less domesticated dialogue between contesting concerns in most other areas where coordination is significantly involved, and this transformation must have as a powerful aim the freeing of time so that the concerns and engagements of receptive generosity might have much broader and deeper influence in our lives. The shape and possibilities of such interventions are far from clear in a world under the sway of a political economy whose increasing power to subjugate people and destroy the earth approaches that of Ockham's God. At this point, as always, the question concerning the openness of history, the possibility of "turning the century with our principles alive," is a practical one. We owe it to Adorno, to ourselves, and to future generations to prove him wrong—to turn the century with our principles alive.

Notes

Introduction

1. See Tzvetan Todorov, *The Conquest of America*, trans. Richard Howard (New York: Harper and Row, 1984).

2. Augustine, *The Confessions*, trans. F. J. Sheed (New York: Sheed and Ward, 1943).

3. E.g., John Milbank, *Theology and Social Theory: Beyond Secular Reason* (Oxford: Basil Blackwell, 1990). For a critical analysis of this work, see my "Storied Others and Possibilities of *Caritas*: Milbank and Neo-Nietzschean Ethics," *Modern Theology* 8, no. 4 (1992).

4. Of course, history is full of examples of hybridization between Christians and non-Christians—from Ireland to South America—which indicate a wild giving and receiving. (From the accounts I am aware of, non-Christians often have been more amenable to this process than Christians.) My hunch is that the better moments within the Christian narratives sometimes overcome the dominant thrust I sketch here.

5. Immanuel Kant, *The Metaphysics of Morals*, trans. Mary Gregor (Cambridge: Cambridge University Press, 1991), 243ff. Cf. Kant, *Lectures on Ethics*, trans. Louis Infield (Indianapolis: Hackett, 1930), 235–36.

6. See Marcel Mauss, *The Gift: The Form and Reason for Exchange in Archaic Societies*, trans. W. D. Halls (New York: W. W. Norton, 1990).

7. I borrow this term from Jacques Derrida, who has used it frequently. See his "Declarations of Independence," trans. Tom Keenan and Tom Pepper, *New Political Science* 15 (1986).

8. The following texts address this issue, some more directly than others: Milbank, *Theology and Social Theory*; Hans Blumenberg, *The Legitimacy of the Modern Age*, trans. R. M. Wallace (Cambridge: MIT Press, 1983); Martin Heidegger, *Nietzsche*, vol. 4: *Nihilism*, trans. F. A. Capuzzi (San Francisco: Harper and Row, 1982); Theodor Adorno and Max Horkheimer, *Dialectic of Enlightenment*, trans. John Cumming (New York: Seabury Press, 1972); Michael Allen Gillespie, *Nihilism before Nietzsche* (Chicago: University of Chicago Press, 1995); Edmund Husserl, *The Crisis of the European Sciences and Transcendental Phenomenology*, trans. David Carr (Evanston: Northwestern

University Press, 1970); Friedrich Nietzsche, *The Will to Power*, trans. Walter Kaufmann and R. J. Hollingdale (New York: Vintage, 1967), bk. 1.

9. In discussing Descartes, Berkeley, and Hume, I do not imply that Newton, Locke, Leibniz, and so forth are unimportant to this story. I am simply guided here by a minimalist heuristic economy.

10. George Berkeley, *A Treatise concerning the Principles of Human Knowledge*, sec. 87, in *The Empiricists* (New York: Doubleday, 1961).

11. Ibid., sec. 86.

12. Ibid., secs. 28 and 29.

13. Ibid., sec. 6.

14. Gerd Buchdahl, *Metaphysics and the Philosophy of Science* (Oxford: Basil Blackwell, 1969), 293.

15. Berkeley, *Treatise*, sec. 140. This idea originates with John Locke's discussion of "Archtypes" in *An Essay concerning Human Understanding* (Oxford: Clarendon Press, 1975).

16. George Berkeley, "Three Dialogues," in *The Empiricists*, 256.

17. David Hume, *A Treatise of Human Nature* (Oxford: Clarendon Press, 1946), 67–68. Cf. Buchdahl, *Metaphysics*, 352.

18. See "On Personal Identity," in Hume's *Treatise*.

19. Immanuel Kant, *Critique of Pure Reason*, trans. Norman Kemp Smith (New York: St. Martin's Press, 1965). Hereafter I will cite the Smith translation, providing the pagination of the German edition (A or B).

20. Friedrich Nietzsche, *Thus Spoke Zarathustra*, trans. Walter Kaufmann (New York: Penguin, 1954), hereafter *TSZ*.

21. Many commentators miss the centrality of this theme in Nietzsche's work. See Kathleen Marie Higgins, "The Night Song's Answer," *International Studies in Philosophy* 17 (Summer 1987), 33–50, and Alexander Nehamas, *Nietzsche: Life as Literature* (Cambridge: Harvard University Press, 1985). Lawrence Lampert's *Nietzsche's Teaching: An Interpretation of "Thus Spoke Zarathustra"* (New Haven: Yale University Press, 1986), a frequently insightful commentary on *Zarathustra*, makes questions concerning gift-giving central. The theme of generosity is present in Bonnie Honig, *Political Theory and the Displacement of Politics* (Ithaca: Cornell University Press, 1993), and in Walter Kaufmann, *Nietzsche: Philosopher, Psychologist, Antichrist* (Princeton: Princeton University Press, 1968); but it is insufficiently developed in the former and poorly developed (through too close an association with Aristotle) in the latter (382–83). While Joseph Beatty focuses his analysis ("Zarathustra: The Paradoxical Ways of the Creator," *Man and World* 3 [1970], 64–75), on giving, he misses most of the profundity of the text by concentrating on the themes of radicalized independence and innocence. Betty tries to be more solar than Zarathustra. My development of the gift-giving virtue takes seriously Hans-Georg Gadamer's emphasis on the importance of the narrative for interpreting the text; see Gadamer, "The Drama of Zarathustra," trans. T. Heilke, in *Nietzsche's New Seas: Explorations in Philosophy, Aesthetics, and Politics*, ed. Michael Gillespie and Tracy Strong (Chicago: University of Chicago Press, 1988).

22. Love and gift-giving distinguish his own teachings from those of his impostures. "Zarathustra's Ape" (*TSZ*, 175–78).

23. Plato, *The Republic*, bk. 7.

24. On the theme of solarity, metaphysics of presence, and Zarathustra, see Jacques Derrida, "Economimesis," *Diacritics* 11, no. 2 (1981), 3–25. I discern a significant stream in *Thus Spoke Zarathustra* that struggles with the problems Derrida identifies.

25. Nietzsche, *The Will to Power*, 232 and 238.

26. Leslie Paul Thiele, in *Friedrich Nietzsche and the Politics of the Soul: A Study of*

Heroic Individualism (Princeton: Princeton University Press, 1990), seems to read Zarathustra as embracing a thoroughgoing solarity from beginning (174) to end (222). This, as I argue below, is to miss some of Nietzsche's most provocative insights (especially concerning relations with others) which Thiele's work does not adequately explore. Tracy Strong, *Friedrich Nietzsche and the Politics of Transfiguration* (Berkeley: University of California Press, 1975), is absolutely right when he points, in contrast to the dominant bent of Thiele and Nehamas, to the great importance Nietzsche places on human interaction.

27. Kathleen Marie Higgins, in *Nietzsche's Zarathustra* (Philadelphia: Temple University Press, 1987), obscures the importance of "Night Song" when she reduces it to a "lament about the emotional strain" of appearing as a "bottomless well of insight and generosity" (136). On the other hand, in "The Night Song's Answer" (1987), her fascinating analysis of Dionysus, Apollo, and Ariadne in regard to the themes of unity, difference, transfiguration, and love, suggests a very fruitful path that the current analysis might explore both to illuminate Nietzsche's sense of Dionysus and to deepen an analysis of receptive generosity. Lampert is correct to read the section as pivotal, the location of "a great shift" (*Nietzsche's Teaching*, 102–5). However, the shift involves a move not just toward the problem of receiving the gift of the doctrine of eternal return, but also—entwined with the former—a move toward the problem of receiving others. Lampert's focus on the growing distance between the philosopher-ruler and other people in *Zarathustra* obscures the numerous, diverse, and important *relations* that Zarathustra both seeks and discovers. See also Nietzsche's discussion in *Ecce Homo*, trans. R. J. Hollingdale (New York: Penguin Books, 1979), pp. 108–110.

28. The centrality of receptivity to giving is insufficiently developed by some of those mentioned above for whom gift-giving is important. Cf. Honig, *Political Theory*, and Kaufmann, *Nietzsche*. Although receptivity is central for Lampert, he ends up conceptualizing it as "letting be"—which overlooks the very important, reciprocally transfigurative and agonistic characteristics of giving and receiving between people in Zarathustra's thinking (developed below).

29. Friedrich Nietzsche, *The Gay Science*, trans. W. Kaufmann (New York: Vintage, 1974), 273–74. There are endless debates on the ontological status of the eternal return in Nietzsche's thought. Kaufmann, *Nietzsche*; Arthur Danto, *Nietzsche as Philosopher* (New York: Macmillan, 1965); and Arnold Zuboff, "Nietzsche and Eternal Recurrence," in *Nietzsche: A Collection of Critical Essays*, ed. Robert Solomon (Garden City, N.Y.: Doubleday, 1973), among others, provide an ontological reading. Lampert (*Nietzsche's Teaching*) views it primarily as a practical regulative idea, as does Nehamas, who provides a compelling discussion of the issue in *Nietzsche*, chap. 5. Nehamas, however, interprets eternal return solely as "a view of the self" (150), a view I wish to decenter through a discussion of receiving otherness "within" and "without," recognizing the real limits of these categories. It is significant that, in *Zarathustra*, the doctrine is articulated not by Zarathustra but by his animals. This distance from Zarathustra suggests that it is a regulative idea, not one he will embrace as an ontological claim, and that we must not reify the idea of eternal return. Zarathustra pokes fun at and doubts his animals when they proclaim the doctrine, pointing perhaps to the fact that it has a certain use but is only one path to the highest virtue. His humor, doubt, and distance open a door to others (like Bernice Johnson Reagon, discussed in my concluding chapter) with other sources which move them in the direction of receptive generosity.

Chapter 1

1. Immanuel Kant, *Critique of Pure Reason*, trans. Norman Kemp Smith (New York: St. Martin's Press, 1965), A126. Hereafter I will cite the Smith translation, providing both the pagination of the German Akademie edition, 3: (A and/or B) and, following a semicolon, the pagination of the translation: e.g., (Axx; 14). For this and other Critiques, the German references are from Immanuel Kant, *Gesammelte Schriften*, 9 vols., ed. Königlich Preussischen Akademie der Wissenschaften (Berlin: Reimer, 1900–1914).

2. I borrow the term "transcendental story" from Henry Allison, *Kant's Transcendental Idealism: An Interpretation and Defense* (New Haven: Yale University Press, 1983).

3. Ibid., esp. 14–34.

4. Ibid., 18–19.

5. Cf. Bxxvi (note).

6. By referring to things in themselves as "absolute otherness," I am not casting my lot resolutely with the vast ranks of those who think that Kant's "thing in itself" denotes another world of entities apart from those which appear. F. H. Jacobi seems to be the earliest representative of this view. Among the many representatives of this position in the twentieth-century literature in English are H. A. Prichard, *Kant's Theory of Knowledge* (Oxford: Clarendon Press, 1909); A. C. Ewing, *A Short Commentary on Kant's Critique of Pure Reason* (Chicago: University of Chicago Press, 1938); and Robert Paul Wolff, *Kant's Theory of Mental Activity* (Gloucester, Mass.: Peter Smith, 1973). Proponents of this view would entangle Kant in a series of contradictions, in which he is said to claim to know of and about a realm he claims he cannot know of and about. Thus, they argue, he delegitimizes knowledge claims concerning that which lies beyond experience and, subsequently, goes on to claim that things exist in this beyond, that they affect our sensibility, and so forth. Though one can find some textual support for this reading, the position it identifies is philosophically so weak as to be hopeless and uninteresting. Moreover, the claim that this reading represents Kant's most persistent and usual understanding of the matter has been rejected, in ways often quite compelling, by scholars such as Henry Allison, Graham Bird, Robert Pippin, Arthur Melnik, and Bernard Rousett, among numerous others. Cf. Allison, *Kant's Transcendental Idealism*, 237–54; Robert Pippin, *Kant's Theory of Form* (New Haven: Yale University Press, 1982), 188–201; and Arthur Melnik, *Kant's Analogies of Experience* (Chicago: University of Chicago Press, 1973), 151–56. The latter three interpreters claim that the distinction between appearances and things in themselves is "methodological," a part of the "transcendental story" which involves two different *ways of considering* objects, rather than two different ontological realms of extant objects. Things in themselves are simply objects "transcendentally considered apart from any way we could know [them]" (Pippin, *Kant's Theory of Form*, 199), and such considerations are merely necessary correlates of the transcendental story about appearances and epistemic conditions. To consider an object apart from our experience of it is not to assert the existence of that object. The merit and intent of such "considerations" lies in their role in helping us understand the necessary "conditions of possibility" of experience, not in claims that there really are objects which correspond to our consideration of things apart from our conditions of knowledge. At his best and most consistent, Kant maintains a critical agnosticism regarding the existence of such things.

While I find this interpretation to be the philosophically most interesting one, Kant's own writings suggest that he had not entirely clarified the issue for himself. For the line of interpretation I'm pursuing, however, either position will do. For even if it were the case that things in themselves refer to an unknowable world of actual entities, the argument I make—about how this absolute otherness calls forth a kind of equivocal receptivity which is a condition of possibility for our sovereignty—would still hold. Its intellectual and aesthetic strength would lie in its illumination of the peculiar relation between self and otherness in the epistemological and ontological dimension of Kantian modernity, and in its account of how Kant was lured (by a yearning for sovereignty) into such incredibly contradictory epistemological and ontological claims. In short, the latter are explained by the paradoxical fact that sovereignty as objective necessity requires a dimension of receptivity but must receive no determinate otherness.

On the other hand, if the "transcendental story" camp is correct, as I think it is for the most part, then my argument can be seen to offer an account of the strange epistemological requirements of Kantian sovereignty which call forth and drive Kant's narrative toward the persistent unfolding of its problematic paradoxes. This account explains why Kant must "think" and write profusely about things in themselves in ways that are utterly at odds with what we can "know." Insofar as Kant was conscious that the transcendental story *is* simply a voyage of considerations and not a voyage into another world of entities, then the course he takes can in no way be attributed to ontological confusion but must be simply a reflection of the imperatives of the subject's desire for a sovereignty of objective necessity itself. My claim is that the doctrine of things in themselves plays a positive role, not only in terms of the regulative ideas of things in themselves introduced in the Transcendental Dialectic (see Pippin, *Kant's Theory of Form*, p. 204–15) but in the central arguments of the Aesthetic and the Analytic as well. Allison makes a parallel argument that Kant "provides a critical justification for the transcendental consideration of affection [and the thing itself thus involved] . . . by the fact that [nonempirical] affection is taken, as a necessary (material) condition of the possibility of experience, and in this sense as a part of a 'transcendental story'" (249). This is certainly true, but its explanatory power seems minimal since the "necessity" Allison refers to is insufficiently elaborated and the problematic equivocations it gives rise to are largely concealed.

7. See H. J. Paton, *Kant's Metaphysic of Experience* (London: George Allen and Unwin, 1936), 1:94.

8. Kant follows Vico in closely associating cognitive grasp and productive mental activity. See *The New Science of Giambattista Vico*, trans. T. G. Bergin and M. H. Fisch (Ithaca: Cornell University Press, 1948).

9. Cf. Allison, *Kant's Transcendental Idealism*, 94–97, and Gerd Buchdahl, *Metaphysics and the Philosophy of Science* (Oxford: Basil Blackwell, 1969), 579, 616. Kant obfuscates his argument when, in the next paragraph, he writes of such determinate qualities as extension and figure as belonging to pure intuition. He specifically acknowledges and remedies this confusion in a later footnote (B160). So, while at times it may appear in this section of Kant's account that the ordering itself comes from sensibility, Kant does eventually make it clear that even pure "space, represented as an object . . . contains more than mere form of intuition; it also contains a *combination* of the manifold. . . . In the aesthetic I have treated this unity as belonging merely to sensibility . . . although, as a matter of fact, it presupposes a synthesis which does not belong to the senses" (B160n). This point is made repeatedly in the deduction in edition A when Kant discusses the three modalities of synthesis.

10. Pippin, *Kant's Theory of Form*, 32.

11. Winfrid Sellars, in *Science and Metaphysics: Variations on Kantian Themes* (London: Routledge and Kegan Paul, 1968), 2ff., also makes this observation, though he draws conclusions markedly different from those I sketch below.

12. Thus, the manifold is similar to what Martin Heidegger calls "standing reserve"

(see "The Question concerning Technology," trans. William Lovitt, in Heidegger, *Basic Writings*, ed. D. F. Krell [New York: Harper and Row, 1977]). Heidegger, however, offers a different reading of Kant in *Kant and the Problem of Metaphysics*, trans. Richard Taft (Bloomington: Indiana University Press, 1990).

13. G. W. F. Hegel, *Phenomenology of Spirit*, trans. A. V. Miller (Oxford: Oxford University Press, 1977), 111–19.

14. While I would not go as far as Wolff's assertion that the Subjective Deduction in the first edition "is the key to the interpretation of the entire *Critique*" (*Kant's Theory of Mental Activity*, 80), I do agree with him that it is here that Kant most extensively "tells us what the word synthesis means. Until we know that we know nothing about the *Critique*" (101). Hence this section significantly colors my thoughts on synthesis, despite Kant's (very problematic) statement that it "does not form an essential part" of his chief purpose (Axvii).

15. Prichard, *Kant's Theory of Knowledge*, 169–70.

16. Ewing, *Short Commentary*, 28, 77.

17. Paton, *Kant's Metaphysic*, 358.

18. Wolff, *Kant's Theory of Mental Activity*, 152–53.

19. See Allison, *Kant's Transcendental Idealism*, 67–68, and W. H. Walsh, *Kant's Criticism of Metaphysics* (Edinburgh: Edinburgh University Press, 1975).

20. Or we must imagine some sort of preestablished legislative harmony of the faculties. To do that, however, would simply transplant precritical, problematic assumptions about the relation between mind and world into the heart of subjectivity without supporting them. See Gilles Deleuze, *Kant's Critical Philosophy*, trans. Hugh Tomlinson and Barbara Habberjam (Minneapolis: University of Minnesota Press, 1984), 22.

21. Kant defines the unthinkable as that which, as self-contradictory, is an impossible thought (Bxxvi). The unthinkability of such givens explains why Kant seldom dwells on them, but rather slips them in, in ways that his interpreters have wrestled with for two hundred years.

22. For a sharp observation related to this point, see Kant's discussion of the second antinomy (A439–44, B467–72).

23. Pippin, *Kant's Theory of Form*, chap. 2, seems to have affinities with some of the argument above. From Paton's contention that Kant didn't really mean that we are given absolute unities, but was just expressing "a limit reached by analysis" (*Kant's Metaphysic*, 358), to Wolff's argument about diverse simultaneities which we cannot know (*Kant's Theory of Mental Activity*, 153–54), to Allison's (following W. H. Walsh) "proleptical" awareness of particulars and organized givens (*Kant's Transcendental Idealism*, 67–68), scores of Kant scholars have tried to resolve this difficulty, but in my view these efforts beg many of the questions raised here.

The most sophisticated attempt to negotiate these difficulties is found in Winfrid Sellars's chapter "Sensibility and Understanding," in *Science and Metaphysics*. Sellars argues that Kant can avoid the problems of the idealist intuiting intellect, on the one hand, and those of the simple impressions of empiricism, on the other, only if he adopts a theory that "analogical counterparts of perception" are given by sensibility and "guide from without" the conceptual synthesis of relations. Without such an analogical view of the relationship between sensibility and understanding, Kant must either posit an "abstract empty" theory of receptivity which can't guide or give anything to categorical synthesis, or he must posit a dimension of determinate order which becomes opaque to synthesis. Sellars argues that sensibility has a "relational structure," "a complex of nonconceptual representations" sufficiently analogous to conceptuality as to guide categorical activity (29–30). Analogy "is obscure and difficult," but "it is nevertheless as essential to the philosophy of science as it has been to theology" (18). "By overlooking the importance of analogical concepts," Kant fell into "empty abstractions" (30). This may be true, but Sellars's account of the happy analogy between sensibility and the understanding rests on a precritical *assumption* of the harmony between the faculties which Kant knew could never ground the subject's sovereignty of objective necessity. In response to Sellars, Kant would have posed something like Deleuze's question: "We have seen that Kant rejected the idea of a pre-established harmony between subject and object. . . . But does he not once again come up with the idea of harmony, simply transposed to the level of faculties of the subject which differ in nature? . . . [T]he Critique demands a principle of the accord" (Deleuze, *Kant's Critical Philosophy*, 22). Whether Kant succeeded here is another question.

24. Francis Hutcheson, in *Collected Works of Francis Hutcheson*, ed. Bernhard Fabian (Hildesheim: Georg Olms Verlagsbuchhandlung, 1971).

25. Alasdair MacIntyre, *Whose Justice? Which Rationality?* (Notre Dame: University of Notre Dame Press, 1988), 270. MacIntyre insightfully develops the tensions and complexities in Hutcheson's thinking. He shows that Hutcheson's rejection of reception in the fundamental epistemological register exists along with an acceptance of receptivity to positive law, tradition, revelation, and so forth, in other registers of his thinking.

26. Immanuel Kant, *The Critique of Practical Reason*, trans. L. W. Beck (Indianapolis: Bobbs-Merrill, 1956), Akademie edition (see note 1 above), 5:66; trans. p. 64. Hereafter I will cite the pagination of the Akademie edition and, following a semicolon, the pagination of the Beck translation: e.g., (22; 20).

27. Theorem I (19; 21ff.).

28. See Henry Allison, *Kant's Theory of Freedom* (Cambridge: Cambridge University Press, 1990), 101–3, and H. J. Paton, *The Categorical Imperative: A Study in Kant's Moral Philosophy* (Philadelphia: University of Pennsylvania Press, 1947), 85–87.

29. Immanuel Kant, *Foundations of the Metaphysics of Morals*, trans. L. W. Beck (Indianapolis: Bobbs-Merrill, 1959), Akademie edition 4:399; trans. 15.

30. Ibid., Akademie edition, 4:429ff.; trans. 47ff.

31. See Allison, *Kant's Theory of Freedom*, 67, and Lewis White Beck, *A Commentary on Kant's "Critique of Practical Reason"* (Chicago: University of Chicago Press, 1960), 267–68, 273.

32. Immanuel Kant, *Critique of Judgment*, trans. W. S. Pluhar (Indianapolis: Hackett, 1987), Akademie edition 5:22; trans. 65. Hereafter I will cite the pagination of the Akademie edition and, following a semicolon, the pagination of the Pluhar translation: e.g., (122; 127).

33. It is little wonder then that many, including Beck in his classic *Commentary*, have found the arguments for the highest good, God, and immortality very problematic. Beck tries to show that these themes are divorceable—and must be divorced—from Kant's conception of moral autonomy. But are they divorceable without ignoring the questions that Kant faced with such integrity? Or do contemporary theorists of sovereign autonomy, of whom Beck is emblematic, end up with a moral position grounded more on finesse than on unconditional reason? Beck's rejection of Kant's conception of the highest good boils down to the following: Kant is "unwilling to draw this conclusion [that moral law alone is the determining ground of the will] in its full force," and he "defiles" the purity of the will with an object that exceeds willful obedience to moral law. Kant does so because he thinks that the moral law "must have an object as well as a form" and thus that the highest good, including happiness, is crucial for the moral disposition of a finite being. Nevertheless, in introducing the highest good, Kant contradicts "what he said earlier and more consistently about the

lawful form of the maxim itself being the object of the moral will" (243). According to Beck, Kant's more reasonable position is that, independent of practical feasibility and inclination, "when deciding on an object . . . we consider only the possibility of rationally willing it . . . if reason completely determined it." In this sense, good "refers only to actions, the maxims that lead to them, the will that produces them." Hence, "the form and the object . . . of the maxim coincide. The object of pure practical reason is not an effect of action but the action itself; the good will has itself as an object" (134). And so, the notion of a necessary object—the highest good—that is more than the good will sacrifices autonomy. Beck claims that Kant's "settled views" are cognizant of this fact, as is indicated by the omission of the highest good from all formulations of the categorical imperative. In short, a command of reason to realize the highest good simply "does not exist" if that command means anything more than "Do your duty" (244–45). Kant introduces it merely for "architectonic purposes of reason" (245), or perhaps as a rationalization of Christian doctrine (270), but it is utterly unnecessary for "a law which speaks to us with commanding authority long before its credentials are presented" (245).

Yet Beck's claim that "the form and the object . . . of the maxim coincide" is problematic, as we have seen. While the form of the moral law is the determining ground for its object, there is no simple identity between the two, as Kant's discussion of the typic makes clear. We have seen above how the highest good functions analogously as an object through which the unconditioned law articulates its unconditionedness. The highest good is the object that the moral law must give itself in order to accomplish the transcendental narrative which at once articulates and further legitimates its receptionless sovereignty (by providing the unconditioned law with positive possession of itself in this object), and this object looms in height and scope superior to all else. No doubt, Beck is right to claim that Kant views moral law as commanding prior to the presentations of its credentials. But this largely misses Kant's point, which is to ground the moral law and make it intelligible in a compelling narrative that secures it against the erosive misconstruals of his philosophical predecessors. The highest good, God, and immortality are crucial to this task. Kant does not, as Beck says he does, "surrender autonomy" by linking moral law and the highest good; rather, he attempts to *salvage* autonomy by giving it an unconditional object instead of a decidedly conditioned object. The latter would itself surrender autonomy through an unconvincing articulation of the unconditioned rational will. All of this indicates, I think, the untenability of these narratives of unconditional, self-given sovereignty as the determining ground of morality. Even as Kant's version fails, it shows Beck's version—insofar as it denies the highest good—to be based on a notion of autonomy that accepts, indeed incorporates in reified form, a highly conditioned finitude into its very being, even as it denies its finitude. By limiting what reason can give itself, reason's claim to be autonomous and sovereign over all else becomes ambiguous, contestable in its essence. In this context, Beck's version becomes only one of several contested conceptions of rationality. As such, it could establish itself (if at all) only on the basis of a difficult debate about the various *conditioned* senses and nonsenses of itself and possible alternatives.

34. Hegel was the first to level this critique—perhaps in "The Spirit of Christianity," but clearly in *Phenomenology of Spirit*, pars. 429–37; *Philosophy of Right*, trans. T. M. Knox (Oxford: Oxford University Press, 1967) sec. 135; and in other texts as well. See Allen Wood, "The Emptiness of the Moral Will," *Monist* 73 (1989), 454–83, for an insightful discussion of the role of this claim in Hegel's thinking. Onora O'Neill, in "Kant after Virtue," *Inquiry* 26 (1983), 387–405, surveys these claims (in the light of Alasdair MacIntyre's criticisms) and argues that, in fact, Kant offers an

ethics of virtues, not an ethics of rules. On O'Neill's reading, Kant was greatly concerned with "the possibility of bringing reason to bear on what is most evidently particular and local about proposals for action" (393). Cf. John Rawls's comment that "in isolation these notions (such as 'a respect for persons') play no role that fixes or limits their use" ("Kantian Constructivism in Moral Theory," *Journal of Philosophy* 77, no. 9 [1980], 517).

35. Alasdair MacIntyre, in *After Virtue*, exemplifies the former possibility; O'Neill, in "Kant after Virtue," the latter.

36. To mention just a few examples: O'Neill, "Kant after Virtue," 390–91; H. B. Acton, *Kant's Moral Philosophy* (London: Macmillan, 1970), 64–65; Robert Paul Wolff, *The Autonomy of Reason* (Gloucester, Mass.: Peter Smith, 1973), 30.

37. For example, Onora Nell [= O'Neill], in *Acting on Principle: An Essay on Kantian Ethics* (New York: Columbia University Press, 1975), iii, distances herself from Kant's examples and treats the question of justification as one that can be answered later if the categorical imperative can be shown to "help us act." The question of justification is addressed thoughtfully in O'Neill's more recent "The Public Use of Reason," *Political Theory* 14, no. 4 (1986), 523–51. However, I believe significant equivocations remain concerning the relation between communication and reason.

38. Kant, *Foundations*, Akademie 4:389; trans. 5.

39. Kant, *Critique of Practical Reason* (36; 38).

40. Kant, *Foundations*, Akademie 4:403; trans. 19.

41. Ibid., Akademie 4:422; trans. 40.

42. Wolff, *The Autonomy of Reason*, 166.

43. Immanuel Kant, "On the Supposed Right to Lie from Altruistic Motives," in Kant, *"Critique of Practical Reason" and Other Writings in Moral Philosophy*, ed. L. W. Beck (Chicago: University of Chicago Press, 1949), 347 (pagination cited hereafter in text).

44. In Immanuel Kant, *"Perpetual Peace" and Other Essays*, ed. Ted Humphrey (Indianapolis: Hackett, 1983), 139.

45. In *Kant: Selections*, ed. L. W. Beck (New York: Macmillan, 1988), 470 n. Beck is wrong insofar as Kant's detailed discussion of unexpected consequences in this case is intended simply to illustrate the indefinite morass one enters by abandoning duty and is not intended as the proof of duty.

46. Acton, *Kant's Moral Philosophy*, 65.

47. Paton, *The Categorical Imperative*, 77 n. 3.

48. See William James Booth, *Interpreting the World: Kant's Philosophy of History and Politics* (Toronto: University of Toronto Press, 1986), 150–53, for this typology and discussion.

49. For some examples, see the following: Howard Williams, *Kant's Political Philosophy* (New York: St. Martin's Press, 1983), 202–6, an attempt to make room for disobedience; P. P. Nicholson, "Kant on the Duty Never to Resist the Sovereign," *Ethics* 77 (1976), 211–33, interpreting the "right of necessity" as an escape hatch in extreme cases; Kenneth Baynes, *The Normative Grounds of Social Criticism: Kant, Rawls, Habermas* (Albany: SUNY Press, 1992), 45–46, claiming that Kant's position here was substantially influenced by his "appeal to a teleological conception of history"; Hans Reiss, "Postscript," in *Kant: Political Writings*, ed. Hans Reiss (Cambridge: Cambridge University Press, 1970), 264–65, arguing that on Kant's grounds, when a system of right ceases, human beings return to a state of nature and therefore may resist an extremely evil government.

50. Kant, *Perpetual Peace* (see note 44 above), 136. This is one reason why Kant argues that the maxim of revolt cannot be publicly declared and hence fails the test of

publicity. See Kant, *Metaphysics of Morals*, trans. Mary Gregor (Cambridge: Cambridge University Press, 1991), Akademie edition 6:319–20; trans. 130–31 (dual pagination cited hereafter in text).

51. See L. W. Beck, "Kant and the Right of Revolution," *Journal of the History of Ideas* 32 (1971), 422.

52. See Williams, *Kant's Political Philosophy*, 183, for an analogy between Kant's categories and the social contract.

53. Hence it is wrong to attribute Kant's conservatism primarily to prudence, as Williams does.

54. For an insightful reading of closures toward otherness, particularly in Kant's *Lectures on Ethics*, trans. Louis Infield (Indianapolis: Hackett, 1963), see Bonnie Honig, *Political Theory and the Displacement of Politics* (Ithaca: Cornell University Press, 1993).

55. There are also crosscurrents in the first two Critiques, but they are not as immediately suggestive with respect to the current discussion.

56. Donald W. Crawford, in "Kant's Theory of Creative Imagination," in *Essays in Kant's Aesthetics*, ed. Ted Cohen and Paul Guyer (Chicago: University of Chicago Press, 1982), favorably views Kant's discussion as continuous with the theme of humanist autonomy. Jacques Derrida, "Economimesis," *Diacritics* 11, no. 2 (1981), 3–25, and Richard Klein, "Kant's Sunshine," *Diacritics* 11, no. 2 (1981), 26–41, both develop a deconstructive reading in which genius is par for the course of Western metaphysics. Jean-François Lyotard, in *The Postmodern Condition*, *Just Gaming*, and *The Differend*, has consistently developed a wild reading of the sublime.

57. Tracy B. Strong and Frank Andreas Sposito's essay "Habermas's Significant Other," in *The Cambridge Companion to Habermas*, ed. S. White (Cambridge: Cambridge University Press, 1995), was published after I completed the manuscript for the present book. In that essay, Strong and Sposito briefly develop Kant's idea of genius as the other of comprehensibility in ways that resonate with some of my following analysis.

58. See Edmund Burke, *A Philosophical Enquiry into the Origin of Our Ideas of the Sublime and the Beautiful* (Oxford: Oxford University Press, 1990). Steven White, in *Edmund Burke: Modernity, Politics, and Aesthetics* (London: Sage, 1994), accents and develops this theme in more ethical and political directions.

59. See Paul Crowther, *The Kantian Sublime: From Morality to Art* (Oxford: Oxford University Press, 1989), 96–102. However, Crowther does not go so far as to question the ways in which such a reading calls into doubt the hegemony of reason.

60. Of course, this tragic sensibility is quickly (but, for reasons stated above, falsely) subdued within the story of the rational vocation of the mind.

61. The universalist implications of Adorno's thinking draw on this idea—but also, and more importantly, on the idea that giving and receiving can be profound even when the possibility of genius is absent from one of the parties (e.g., someone in a coma, near death). Adorno is animated partly by the tragic in a way that Kant is less attuned to here.

62. Friedrich Nietzsche, *Beyond Good and Evil*, trans. Walter Kaufmann (New York: Vintage Books, 1966), sec. 274.

63. In *Anthropology from a Pragmatic Point of View*, trans. Mary Gregor (The Hague: Nijhoff, 1974), Kant writes: "The man of genius cannot explain to himself its outburst or how he arrived at a skill which he never tried to learn" (sec. 58).

64. See Hans-Georg Gadamer, *Truth and Method*, 2d ed. trans. Joel Weinsheimer and D. G. Marshall (New York: Crossroad Press, 1989), 49–60. Also see Gadamer's illuminating discussion of the "subjectivization of aesthetics" (42–81) for a critical reading of subjectivism in the *Critique of Judgment* that significantly overlaps in orientation with my reading of the first two Critiques.

65. See Derrida, "Economimesis," for a critical reading, and Crowther, *The Kantian Sublime*, esp. chap. 7, for a reading essentially affirming what I am calling "solarity."

66. Or, seeking to avoid Kant's reifying theory of the faculties, one might speak of relations of reciprocal agonistic provocation among various levels of, and strivings for, determinacy and comprehensiveness in the "clearing" of an embodied being.

67. This reading enables us to locate the otherwise missing discussion of an artistic sublime in the aesthetic idea. See Crowther, *The Kantian Sublime*, chap. 7.

68. For a definitive rejection of the theory of inquiry employed by Kant, see Thomas Kuhn's *The Structure of Scientific Revolutions* (Chicago: University of Chicago Press, 1962).

69. Immanuel Kant, "An Answer to the Question: What Is Enlightenment?" in Kant, *"Perpetual Peace" and Other Essays*, Akademie edition 8:35; trans. 41. Hereafter I will cite the pagination of the Akademie edition and, following a semicolon, the pagination of the Humphrey translation: e.g., (36; 41–42).

70. For an entirely *sovereign* interpretation of active enlightenment, see the *Critique of Judgment* (294–95; 161).

71. In *Kant: Political Writings* (see note 49 above), 248.

Chapter 2

1. See "Jürgen Habermas: A Generation Apart from Adorno" (interview), *Philosophy and Social Criticism* 18, no. 2 (1992), 122. Peter U. Hohendahl, in "The Dialectic of Enlightenment Revisited: Habermas' Critique of the Frankfurt School," *New German Critique* 35 (Spring–Summer 1985), 3–26, provides a careful and illuminating account of the development of Habermas's relation to Adorno from the early 1960s forward.

2. Other theorists follow the contours of this critique quite closely. See, for example, Seyla Benhabib, *Critique, Norm, and Utopia: A Study of the Foundations of Critical Theory* (New York: Columbia University Press, 1986); Axel Honneth, "Communication and Reconciliation: Habermas's Critique of Adorno," *Telos* 39 (Spring 1979), and *The Critique of Power: Reflective Stages in Critical Social Theory*, trans. K. Baynes (Cambridge: MIT Press, 1991); Thomas McCarthy, *The Critical Theory of Jürgen Habermas* (Cambridge: MIT Press, 1978); and Albrecht Wellmer, *The Persistence of Modernity: Essays on Aesthetics, Ethics, and Postmodernism*, trans. David Midgley (Cambridge: MIT Press, 1991), though Wellmer's critique deviates somewhat and is more subtle. There are resonances between Habermas's critique and Leszek Kolakowski's far more damning analysis: "There can be few works of philosophy that give such an overpowering impression of sterility as *Negative Dialectics*" (Kolakowski, *Main Currents of Marxism*, vol. 3; *The Breakdown* [Oxford: Oxford University Press, 1970], 366). Similarly, Stanley Rosen, *Hermeneutics and Politics* (Oxford: Oxford University Press, 1987), 10, writes: "It is obvious that negative dialectics will not help."

3. I do not have space here to discuss this reception in full detail. Jean-François Lyotard, in "Adorno as Devil," *Telos* 19 (Spring 1974), 127–37, provides one of the least insightful readings when he writes that, in Adorno, "the category of the subject remains uncriticized" (127). Cf. Michael Ryan, *Marxism and Deconstruction: A Critical Articulation* (Baltimore: Johns Hopkins University Press, 1982), and Rainer Nagele, "The Scene of the Other: Theodor W. Adorno's Negative Dialectics in the Context of Poststructuralism," in *Postmodernism and Politics*, ed. Jonathan Arac (Minneapolis: University of Minnesota Press, 1986). Shane Phelan, in "Interpretation and Domination:

Adorno and the Habermas–Lyotard Debate," *Polity* 25, no. 4 (1993), 597–616, offers a very sympathetic and insightful reading of the issue at hand.

4. Michael Shapiro, "Politicizing Ulysses: Rationalistic, Critical, and Genealogical Commentaries," *Political Theory* 17 (February 1989), 21–22.

5. See Stanley Fish, *Doing What Comes Naturally* (Durham: Duke University Press, 1989), 455–57.

6. At least this would seem necessary for all activity that exceeds the anonymous flow of herdlike practice or oblivious decisionism.

7. Other efforts to draw out the ethical implications of Adorno's work include Robert Hullot-Kentor, "Back to Adorno," *Telos* 81 (Fall 1989), 5–29; Drucilla Cornell, "The Ethical Message of Negative Dialectics," *Social Concept* 4 (1987), 30–36; and Fred Dallmayr, *Twilight of Subjectivity: Contributions to a Post-Individualist Theory of Politics* (Amherst: University of Massachusetts Press, 1981) and *Between Freiburg and Frankfurt: Toward Critical Ontology* (Amherst: University of Massachusetts Press, 1991). Although Cornell and Dallmayr insightfully explore receptivity and difference in Adorno, their interpretations place less emphasis on the agonistic dimension of his work; moreover, I think Dallmayr's reading may have to go further toward addressing some of Richard Bernstein's questions about universality (see Bernstein, "Fred Dallmayr's Critique of Habermas," *Political Theory* 16, no. 4 [1988], 580–93).

8. Susan Buck-Morss's masterful work *The Origin of Negative Dialectics: Theodor W. Adorno, Walter Benjamin, and the Frankfurt Institute* (New York: Free Press, 1977), esp. chaps. 5 and 6, traces Adorno's work on constellations to Walter Benjamin. See Benjamin's "Epistemo-Critical Prologue" (1924–25), in *The Origin of German Tragic Drama*, trans. John Osborne (London: New Left Books, 1977), the essay in which the term is first employed. Adorno taught Benjamin's text in 1931, and his essay of that year, "The Actuality of Philosophy," *Telos* 31 (Spring 1977), 120–31, explicitly utilized and transfigured Benjamin's metaphor. The concept continues to change up until the writing of *Negative Dialectics*, where Adorno's relation to science, questionableness, and "keys versus combinations" is substantially different from his earliest articulations. See also part 1 in Fredric Jameson, *Late Marxism: Adorno; or, The Persistence of the Dialectic* (London: Verso, 1990); Gillian Rose, *The Melancholy Science* (New York: Columbia University Press, 1978); Lambert Zuidervaart, *Adorno's Aesthetic Theory: The Redemption of Illusion* (Cambridge: MIT Press, 1991), 60–64; and Phelan, "Interpretation and Domination."

9. Perhaps the real force of this point has been substantially missed by the interpreters mentioned in note 8 above.

10. This argument is echoed and developed by Honneth, "Communication and Reconciliation."

11. This critique is articulated at length in Adorno's *Against Epistemology: A Metacritique*, trans. W. Domingo (Cambridge: MIT Press, 1983).

12. Seyla Benhabib follows this line of analysis for the most part, though she thinks there was "a tension, if not an incompatibility" between Horkheimer, who she argues maintained a belief that reason had an emancipatory force, and Adorno, who "views reason to be inherently an instrument of domination." See Benhabib, *Critique, Norm, and Utopia*, 163ff. Habermas himself makes a similar distinction between the two authors in "Nachwort von Jürgen Habermas," in Max Horkheimer and Theodor Adorno, *Dialektik der Aufklärung* (Frankfurt: Fischer, 1986). See Hullot-Kentor, "Back to Adorno," for a penetrating critique of Habermas's essay.

13. Theodor Adorno, "A Portrait of Walter Benjamin," in *Prisms*, trans. Samuel and Shierry Weber (Cambridge: MIT Press, 1982), 234.

14. Hans Blumenberg, in *Legitimacy of the Modern Age*, trans. Robert Wallace

(Cambridge: MIT Press, 1983), outlines the debate on modernity's continuity versus discontinuity with the past, arguing for the latter.

15. The idea of "stages" seems to be based on a number of very problematic anthropological and historical assumptions, more dominant in the 1940s than today. In a related fashion, Adorno and Horkheimer frequently refer to "primitive societies" in a singular totalizing way when, in fact, the multitudinous societies thus subsumed vary greatly from one another. My appropriation jettisons this totalizing aspect.

16. Theodor Adorno and Max Horkheimer, *Dialektik der Aufklärung: Philosophische Fragmente*, in Adorno, *Werke*, 20 vols., ed. Rolf von Tiedeman (Frankfurt am Main: Suhrkamp, 1981), 3:31–32. All other German references are to this edition. Benhabib draws heavily on this passage in order to support an interpretation of the *Dialectic of Enlightenment* that is very much at odds with the one I am developing here. Needless to say, I think she misreads it. Interestingly, the line on the movement of language from tautology to language (expressing a dialectical tension between identity and nonidentity, distinctness and relation), whereby thinking in a more desirable sense becomes possible, is omitted from Benhabib's lengthy quotation on p. 168 of *Critique, Norm, and Utopia*. So, too, she omits the line which claims that it is *owing to the structure of terror and taboo in the face of otherness* that this dialectic remains impotent, implying again that beyond this taboo is a certain more desirable possibility. By omitting these passages, it is possible—but still, I think, by no means easy—to read the remaining lines such that they support her thesis concerning the text's self-defeating aporetical structure.

17. Again I find myself in disagreement (for reasons that should be apparent below) with Benhabib's interpretation of this chapter. She argues that of the two excursuses in the *Dialectic of Enlightenment*, the one on the *Odyssey* contains a much stronger condemnation of Western reason as irredeemably subjugative, and that the one labeled "Enlightenment and Morality" contains a more measured criticism. Adorno is said to be responsible for the darker reading, while Horkheimer was supposedly the primary author of the more hopeful excursus. Habermas makes a similar claim in his "Bermerkungen zur Entwicklung des Horkheimerischen Werkes," cited in Hullot-Kentor, "Back to Adorno," 5–28. Hullot-Kentor provides a compelling refutation of this and many other frequently reiterated Habermasian readings of Adorno.

18. Joel Whitebook, frequently an insightful critical theorist, reads this passage to imply that "all development is, *a priori*, self-defeating" ("Reason and Happiness: Some Psychoanalytic Themes in Critical Theory," in *Habermas and Modernity*, ed. Richard Bernstein [Cambridge: MIT Press, 1985], 146). I counter this reading extensively when I bring Adorno to bear critically on Habermas's understanding of inner nature.

19. I have relied here upon Robert Hullot-Kentor's excellent and much needed translation of the Odysseus chapter in *New German Critique* no. 56 (1992), 139. Cf. *DE*, 76.

20. We shall see below that this longing is only one point in the tension-filled constellation that guides Adorno's thinking—a constellation that leads him in directions very different from the longing for total revolution, absolute freedom, and associated terror that it might engender if considered as a singular absolute imperative. Bernard Yack, in *The Longing for Total Revolution: Philosophical Sources of Social Discontent from Rousseau to Marx and Nietzsche* (Princeton: Princeton University Press, 1986), misreads Adorno on this point (see 286). While I agree with Yack on the dangers of this longing, I argue below (reading Adorno) that we should not jettison transcendent longings, but should juxtapose them with radically opposed insights in order to compose a constellation of ethical solicitations and illuminations that might animate more receptive and generous engagements with others and with the nonhuman world. See

the concluding chapter in Charles Taylor, *Hegel* (Cambridge: Cambridge University Press, 1975), for a very compelling interpretation of Hegel's critique of absolute freedom.

21. Hullot-Kentor trans., 140; cf. *DE*, 78.

22. Hullot-Kentor trans., 140; cf. *DE*, 79.

23. Walter Benjamin, in his "Theses on the Philosophy of History," in *Illuminations*, trans. Harry Zohn, ed. Hannah Arendt (New York: Schocken, 1969), gives this theme its most powerful articulation.

24. See Emmanuel Levinas, *Totality and Infinity: An Essay on Exteriority*, trans. Alfonso Lingis (Pittsburgh: Duquesne University Press, 1969), 42–48, where ethics and metaphysics appear to be interchangeable terms.

25. Some of the soliciting dimensions of aesthetics (and, I would argue, of thought as well) are already evoked in Adorno's *Kierkegaard: Construction of the Aesthetic*, trans. Robert Hullot-Kentor (Minneapolis: University of Minnesota Press, 1989), when he writes: "It is the cell of a materialism whose vision is focused on 'a better world'—not to forget in dreams the present world, but to change it by the strength of an image" (130–31). Adorno frequently writes of being "surrounded" by images, a thought similar to that of being positioned within a constellation of diversely soliciting images and thoughts. Earlier in this same text, passages quoting Kierkegaard on paradox evoke much of my interpretation of Adorno's later texts: "The paradox is the source of the thinker's passion, and the thinker without a paradox is like a lover without feeling: a paltry mediocrity" (114). The will to paradox, Adorno notes, can itself become abstract and thus negate itself—by short-circuiting the concrete transfiguring engagements with the world for a sacrificial dogma of paradox.

26. It would be a mistake, as we shall see, to assume that on Adorno's reading the aim of thinking is simply—if tragically—correspondence with the world. His point here for the moment is only that, however we may conceive of the aims of thinking with respect to the world that exceeds it, our substantial blindness would seem to pose a problem for all but the most positivistic visions.

27. For the German, see Theodor Adorno, *Negative Dialektik* (Frankfurt: Suhrkamp, 1973), 17. In the English edition, *Bewusstsein* is "sense."

28. Much of Wellmer's *The Persistence of Modernity* addresses these and related issues. Among these essays, the most sustained discussion occurs in "The Dialectic of Modernism and Postmodernism: The Critique of Reason since Adorno," 36–94.

29. Ibid., 71.

30. Ibid., 74.

31. Ibid., 73.

32. Susan Buck-Morss (*The Origin of Negative Dialectics*, esp. 57–62) probingly explores Adorno's transfiguring employment of concepts, as does Gillian Rose (*The Melancholy Science*, chap. 2).

33. See Adorno, "The Actuality of Philosophy," 120–33.

34. No one explores this more profoundly than Maurice Merleau-Ponty. See my discussion in *Self/Power/Other: Political Theory and Dialogical Ethics* (Ithaca: Cornell University Press, 1992), chap. 4.

35. This term comes from Fred Dallmayr, "The Politics of Non-Identity: Adorno, Postmodernism and Edward Said," a paper delivered at the annual meeting of the American Political Science Association, September 1994.

36. This potentially circular and reinforcing relation between the capacity to become conscious of the delusions of consciousness, on the one hand, and an awareness of tragic finitude, on the other, is extremely important. For this relation is a locus of the critical power of negative dialectics. In response to a critic like Stanley Fish, who

claims that "critical self-consciousness is at once impossible and superfluous" (*Doing What Comes Naturally*, 464) because one can gain no distance from one's historical situatedness and (therefore) all situations are equally constraining, Adorno's position is that "how we think about thought" matters with respect to the amount of critical distance we are likely to achieve. An awareness of tragic finitude can animate a questioning-and-distancing relation with our thought and discursive practices that is more likely to discern weakness, blindness, contradiction. And such an awareness is more likely to draw its future identities in ways that are more expansive, variegated, and inclusive, and that simultaneously foreground contingencies, tensions, incompletion, and blindness such that they make greater space for—and even solicit—the interrogative probing of other possibilities, instead of freezing them in the background through strategies that reinforce and reward blind complacency. Of course, infusing these questions of qualitatively different directions are numerous fuzzy questions of degrees of critical awareness, but that does not make negative dialectics impossible or any less important. Fish levels the entire issue by (a) remaining at the most abstract consideration of foreground and background (we always think against a background) and (b) ignoring the ways in which foregrounded discursive practices powerfully contribute to the definition and structure of our relation to submerged background possibilities (a relation which varies widely in terms of inclusiveness, fluidity, solicitiveness, and a tragic sensibility that animates the opening toward otherness).

37. By which Adorno means a situation where giving is increasingly relegated to bureaucratic charities that facilitate the "planned papering-over of society's visible sores" while they often humiliate and objectify the recipients; a situation in which the rule of exchange value and anonymity in gift-giving is exemplified both by the proliferation of generic "gift-articles, based on the assumption that one does not know what to give because one really does not want to," and by the "right to exchange, which signifies to the recipient: take this, it's all yours, do what you like with it; if you don't want it, that's all the same to me, get something else instead" (*MM*, 43).

38. For the German, see p. 66 of the edition cited in note 37 above. This passage is a difficult one. The first part of the third sentence is particularly difficult to make sense of in light of the flow of Adorno's thinking, which is why it is rendered diversely by translators: "Es ist das Mögliche, nie das unmittelbar Wirkliche, das der Utopie den Platz versperrt . . ." The paradoxes here lead David Held to translate thus: "It is the possible, never the immediately existing, that contains locked up within itself a place for Utopia" (*Introduction to Critical Theory: Horkheimer to Habermas* [Berkeley: University of California Press, 1980], 221). While our interpretations of this particular passage diverge somewhat in letter, I think in terms of the spirit of *Negative Dialectics*, regarding "possibility," we are not so distant.

39. Adorno's treatment of Hegel in the introduction to *Negative Dialectics* illustrates constellational thinking as exemplification. One can sense Adorno dancing around Hegel from many discrepant angles, bringing those angles to bear on one another, rendering subtle changes in light of these tensions, generating new ones. What emerges is not only one of the most powerful and least reductive readings of Hegel, but an exemplification of a thinking engaged in receptive generosity. For a similar observation on Adorno's constellational approach to art, see Zuidervaart, *Adorno's Aesthetic Theory*, 62.

40. Exercising restraint, I forgo discussion of several other points in Adorno's constellation of ethical solicitations. Similar analyses could be articulated regarding agonistic points of illumination such as equality and difference, universality and particularity, seriousness and play, immanence and exteriority, system and antisystem, totality and infinity, thought and rhetoric, nature and history, essence and appearance.

A full-blown effort to articulate what Adorno called the "morality of thinking" would require further exploration of these points.

41. This passage contains insights that overlap with some of Friedrich Nietzsche's observations on language in sec. 354 of *The Gay Science*, trans. W. Kaufmann (New York: Vintage, 1974).

42. See Max Horkheimer, "Means and Ends," in *Eclipse of Reason* (New York: Continuum, 1947).

43. It is interesting to read this analogy as an intensification of, or variation on, the pressures on selves that Hobbes thought were necessary for building a Leviathan: "not unlike to that we see in stones brought together for building of an ædifice," when irregular recalcitrant stones (or humans) "the builders cast away as unprofitable, and troublesome" (Thomas Hobbes, *Leviathan*, ed. C. B. Macpherson [New York: Penguin, 1968], 209).

44. For an exploration of this animating paradox, see Hendrik Birus, "Adorno's 'Negative Aesthetics'?" in *Languages of the Unsayable: The Play of Negativity in Literature and Literary Theory*, ed. S. Budick and W. Iser (New York: Columbia University Press, 1989), 140–64.

45. Adorno thinks that the mimetic moment is greatly accented in art, and less so in philosophy, but that it is present in both. Fredric Jameson, in *Late Marxism*, repeatedly discusses the concept of "mimesis" in Adorno's work in subtle and illuminating ways; see particularly p. 68.

46. Here I borrow and obviously alter a term that is central to Levinas in *Totality and Infinity*, 194ff.

47. Significantly, it is to highlight the "noncommunicative" moment within our dialogical efforts that Adorno not infrequently employs the analogy of "labor and material" to depict the relation between thought and the world (see *ND*, 19). He uses this analogy to capture the quality of silent somatic resistance and reciprocity that is an inextinguishable aspect of the relation between thought and the world, an aspect that infiltrates quite thoroughly all our communicative relations, all our concepts. Accenting this moment, Adorno writes: "Subjectivity comes into its own not as communication or message, but as labor" (*AT*, 238). Yet if this laboring dimension calls us away from the subjectivity of the world, its *telos* is always a reciprocity of expression between self and world that spills beyond the laboring relation and draws us back to efforts to communicate with others, albeit as beings with a voice different from that which we had before our efforts to leave.

48. See *MM*, 105, on the catastrophic dialectic involved in denying the faces of animals.

49. As Albrecht Wellmer tends to do in "Truth, Semblance, and Reconciliation: Adorno's Aesthetic Redemption of Modernity," in *The Persistence of Modernity*. Wellmer focuses too exclusively on "truth context" in Adorno's aesthetics.

50. It should be clear from this passage—as well as from Adorno's frequent comments on the body, intuition, mimetic desire, shock, suffering, happiness, etc.—that Wellmer exaggerates when he claims (ibid.) that Adorno's aesthetic is cognitive to the total exclusion of perception and emotion (though Wellmer's analysis is not without some truth concerning the accent of many of Adorno's reflections on aesthetics). See the discussion of Wellmer's analysis in Zuidervaart, *Adorno's Aesthetic Theory*, chap. 11.

51. Making a different but resonant point, Adorno claims: "There may be a sense in which an extraneous standpoint is necessary, or else art is apt to fetishize its autonomy" (*AT*, 358). Also: "Called to legitimate itself before the world of things, art—a thing negating the thing-world—appears *a priori* helpless. All the same, art can-

not simply shirk the responsibility of legitimating itself in the eyes of the world" (175).

52. In another, related passage it is the religious animal—or some part of it—that Adorno has in mind, echoing Goethe: "Enter into works of art as you would into a chapel" (*AT*, 489).

53. Charles Taylor, *Sources of the Self: The Making of the Modern Self* (Cambridge: Harvard University Press, 1989), 516.

54. Ibid., 517.

Chapter 3

1. Albrecht Wellmer, *The Persistence of Modernity: Essays on Aesthetics, Ethics, and Postmodernism*, trans. D. Midgley (Cambridge: MIT Press, 1991), 183–84.

2. David Ingram views this argument as evidence that Habermas is "expanding communicative rationality . . . to include, in however subordinate a role, a poetic moment of deconstruction" (*Habermas and the Dialectic of Reason* [New Haven: Yale University Press, 1987], 92). My reading is diametrically opposed to this, unless one includes total subjugation as a mode of "inclusion."

3. See Karl O. Apel, "The A Priori of the Communication Community and the Foundations of Ethics," in Apel, *Towards a Transformation of Philosophy* (London: Routledge and Kegan Paul, 1980).

4. Jean-François Lyotard levels this criticism at Habermas in a not very subtle fashion in *The Postmodern Condition: A Report on the Condition of Knowledge*, trans. G. B. Massumi (Minneapolis: University of Minnesota Press, 1984). See Richard Rorty's response in "Habermas and Lyotard on Postmodernity," in *Habermas and Modernity*, ed. Richard Bernstein (Cambridge: MIT Press, 1985). Fred Dallmayr develops this same critique more insightfully in *Between Freiburg and Frankfurt: Toward Critical Ontology* (Amherst: University of Massachusetts Press, 1991), and Thomas Dumm makes the point in "The Politics of Post-Modern Aesthetics: Habermas contra Foucault," *Political Theory* 16, no. 2 (1988), 209–28.

5. Jürgen Habermas, "A Reply to My Critics," in *Habermas: Critical Debates*, ed. John Thompson and David Held (Cambridge: MIT Press, 1982), 235, hereafter "RC."

6. See *PDM*, chap. 9, and Michel Foucault, *The Order of Things: An Archaeology of the Human Sciences* (New York: Vintage, 1973).

7. Habermas's earliest formulation of this position was in a 1968 essay, in "Technology and Science as 'Ideology,'" reprinted in Habermas, *Toward a Rational Society*, trans. Jeremy Shapiro (Boston: Beacon Press, 1973). In that essay, Habermas focused his critique on Herbert Marcuse, whose understanding of reconciliation with nature in *One Dimensional Man* (Boston: Beacon Press, 1964) is very different from Adorno's. C. Fred Alford, in *Science and the Revenge of Nature* (Gainesville: University Presses of Florida, 1985), discusses the Marcuse–Habermas difference at length.

8. Jürgen Habermas, *Communication and the Evolution of Society*, trans. Thomas McCarthy (Boston: Beacon Press, 1979), 93.

9. See David Rasmussen, *Reading Habermas* (Cambridge, Mass.: Basil Blackwell, 1990), for an extensive examination of the transcendental aspect of Habermas's project.

10. Joel Whitebook, "The Problem of Nature in Habermas," *Telos* 40 (Summer 1979), 50–51.

11. Jürgen Habermas, "Questions and Counterquestions," in *Habermas and Modernity*, ed. Richard Bernstein (Cambridge: MIT Press, 1985), 213, hereafter "QC."

12. Joel Whitebook, "Reason and Happiness: Some Psychoanalytic Themes in Critical Theory," in *Habermas and Modernity*, 156.

13. Obviously, Freud's *Civilization and Its Discontents* stands in marked contrast to Habermas here.

14. Here I am at odds with Whitebook's interpretation of "Adorno's wholesale appropriation of id psychology," in "Reason and Happiness," 144. Adorno's position is far more subtle and interesting than Whitebook recognizes. In a recently published book, Whitebook reiterates the same position, following Habermas's and Wellmer's critiques of Adorno at length. See Joel Whitebook, *Perversion and Utopia: A Study in Psychoanalysis and Critical Theory* (Cambridge: MIT Press, 1995). In the last several pages of his book, Whitebook begins an alternative reading of "the opposing tendencies of [Adorno's] thinking" that does more justice to Adorno's work than the primary reading Whitebook offers. This alternative is evoked (in a passage Whitebook quotes in the final pages) when Adorno writes of a relation between consciousness and the somatic—i.e., beyond the rigid opposition of conscious and unconscious—that is preferable to the simplistic understandings: "But if the impulses are not at once preserved and surpassed in the thought which has escaped their sway, then there will be no knowledge at all, and the thought that murders the wish that fathered it will be overtaken by the revenge of stupidity" (*MM*, 122). This thought infuses every word Adorno wrote, not just his *Aesthetic Theory*. Yet, while Adorno clearly thinks there is a desirable, expressive, and hopeful aspect of *Aufhebung*, he does not think that *Aufhebung* is ever accomplished without some tragedy. This doesn't lead to a totalizing damnation of synthesis, but it does engender moments of skepticism, irony, and ongoing agonistic interrogations. And it leads to a different understanding of "the vitality of the self," as Whitebook phrases it.

Whitebook's interpretation of Hans Leowald contains much that is fascinating (and akin to much in Adorno) concerning the intricacies of sublimation (and of pleasure) that can account for sublimation's compelling and desirable aspects. Because Whitebook remains within the Habermasian discursive economy concerning Adorno, however, Adorno is never permitted to interrogate Leowald in ways that I think are crucial. Thus, from Adorno's perspective, Leowald/Whitebook's view that "the vitality of the self is generally proportional to the extent it has integrated the disavowed, split-off, and heterogeneous parts of the psyche" (*Perversion and Utopia*, 257) is very problematic. Adorno would say, more wisely I think, that the vitality of selves is proportional to the extent to which we maintain a tension between cultivating (a) our integrative impulses and achievements and (b) our active sense of, and endless engagements with, that which resists thorough integration and is pregnant with possibilities that lie beyond any extant synthesis. "Life that lives" is life that dwells in the tension between these two points. Thus Adorno (who also cautioned against fetishizing paradox in a way that would cease to be paradoxical, i.e. cease to throw us into engagement with nonidentity) quotes Kierkegaard: "The paradox is the source of the thinker's passion" (Theodor Adorno, *Kierkegaard: Construction of the Aesthetic*, trans. Robert Hullot-Kentor [Minneapolis: University of Minnesota Press, 1989], 114). The paradox is also vital to taming "the pernicious tendency towards all-pervasive . . . subsumption" linked to the authoritarian personality. See Theodor Adorno et al., *The Authoritarian Personality* (New York: W. W. Norton, 1982), 349. Jessica Benjamin, in "The End of Internalization: Adorno's Social Psychology," *Telos* 32 (Summer 1977), 42–63, misreads Adorno in a similar way, and also fails to note the intersubjective dimension of his thinking.

15. See Romand Coles, "Ecotones and Environmental Ethics: Adorno and Lopez," in *In the Nature of Things*, ed. J. Bennett and W. Chaloupka (Minneapolis: University of Minnesota Press, 1993), 226–49.

16. Donna Haraway, "Situated Knowledges: The Science Question in Feminism and the Privilege of Partial Perspective," *Feminist Studies* 14 (1988), 595. Also see Jane Bennett's provocative discussion in "Primate Visions and Alter-Tales," in *In the Nature of Things* (ibid.).

17. See Nicholas Xenos, *Scarcity and Modernity* (London: Routledge, 1989), for a provocative analysis of the emergence of the concept of scarcity during the Scottish Enlightenment, as well as explorations of the problems, dangers, and limits of this concept.

18. Jürgen Habermas, "Walter Benjamin: Consciousness-Raising or Rescuing Critique," in *Philosophical-Political Profiles*, trans. F. G. Lawrence (Cambridge: MIT Press, 1983), 131–66.

19. For a reading which pushes on these openings, see Stephen White, "Foucault's Challenge to Critical Theory," *American Political Science Review* 80, no. 2 (1986), and *The Recent Work of Jürgen Habermas: Reason, Justice, and Modernity* (Cambridge: Cambridge University Press, 1988). In these texts, though, White is insufficiently cognizant of the insistences and closures that operate within Habermas's work, including the sharp separation between the aesthetic and the moral realm. For a markedly different reading of the contrasts between Foucault and Habermas, see Dumm, "The Politics of Post-Modern Aesthetics"; Dumm's question to White (225 n. 17) is an extremely important one concerning the assumption of a radical division between the juridical and the aesthetic. White engages postmodern concerns with greater illumination and sensitivity in *Political Theory and Postmodernism* (Cambridge: Cambridge University Press, 1991), but he still maintains too sharp a disjuncture between a "responsibility to act" and a "responsibility to otherness." I argue ("Identity and Difference in the Ethical Positions of Adorno and Habermas," in *The Cambridge Companion to Habermas*, ed. S. White [Cambridge: Cambridge University Press, 1995], 37–38) that this disjuncture is flawed to the extent that each responsibility is meaningless if it does not contain the other within itself. Dumm's reading of Habermas, while illuminating some important weaknesses, too polemically eclipses the more "resistant" aspects of Habermas's theorizing; and Dumm fails to subject his own orientation to the very kind of questioning that he criticizes Habermas for avoiding. Cf. my "Communicative Action and Dialogical Ethics: Habermas and Foucault," *Polity* 25, no. 1 (1992).

20. Here, as in many places, Adorno has substantial affinities with Jacques Derrida, who has written since the early 1960s about giving beyond the law. For Derrida's recent thoughts on this theme, see Derrida, *Given Time: Counterfeit Money*, trans. P. Kamuf (Chicago: University of Chicago Press, 1992).

21. For an insightful development of this concept, see William E. Connolly, *Politics and Ambiguity* (Madison: University of Wisconsin Press, 1987).

Conclusion

1. Leszek Kolakowski, *Main Currents of Marxism*, vol. 1: *The Founders* (Oxford: Oxford University Press, 1978).

2. Georg Lukács, *History and Class Consciousness* (Cambridge: MIT Press, 1971), is one of the most insightful Marxian accounts of these relations. For the importance of Fichte, see Tom Rockmore, *Fichte, Marx, and the German Philosophical Tradition* (Carbondale: Southern Illinois University Press, 1980). Cf. Bernard Yack's *The Longing for*

Total Revolution: Philosophical Sources of Social Discontent from Rousseau to Marx and Nietzsche (Princeton: Princeton University Press, 1986).

3. In *The Marx-Engels Reader*, 2d ed., ed. Robert Tucker (New York: W. W. Norton, 1978), 54, hereafter *MER*.

4. Immanuel Kant, *Critique of Pure Reason*, trans. Norman Kemp Smith (New York: St. Martin's Press, 1965), Bxxiii.

5. William James Booth, in "Gone Fishing: Making Sense of Marx's Concept of Communism," *Political Theory* 17, no. 2 (1989), insightfully explores questions of time and political economy in Marx's understanding of capitalism and communism.

6. But they didn't, according to Marx. Marx's essay, written immediately after the defeat of the Paris Commune, is a laudatory obituary for a revolt of which—as his letters reveal—he was in fact quite critical. See *The Letters of Karl Marx*, ed. Saul K. Padover (Englewood Cliffs, New Jersey: Prentice Hall, 1979), 280–82 and 334. For a more detailed account of the Commune and Marx's relation to it, see Isaiah Berlin, *Karl Marx: His Life and Environment* (Oxford: Oxford University Press), chap. 10. The Commune was, like any revolution, full of messy contestation—which raises questions that follow.

7. The problems I explore below are relevant to numerous other theorists who have many affinities to Ernesto Laclau and Chantal Mouffe. Like Laclau and Mouffe, these theorists are highly insightful in many ways, but are weaker concerning the question at hand. See Iris Marion Young, *Justice and the Politics of Difference* (Princeton: Princeton University Press, 1990) and Samuel Bowles and Herbert Gintis, *Capitalism and Democracy: Property, Community, and the Contradictions of Modern Social Thought* (New York: Basic Books, 1987).

8. Chantal Mouffe, "Preface: Democratic Politics Today," in *Dimensions of Radical Democracy: Pluralism, Citizenship, Community*, ed. Chantal Mouffe (London: Verso, 1992), 1.

9. Ernesto Laclau, "Community and Its Paradoxes: Richard Rorty's 'Liberal Utopia,'" in *Community at Loose Ends*, ed. Miami Theory Collective (Minneapolis: University of Minnesota Press, 1991), 97. Cf. Richard Rorty, *Contingency, Irony, Solidarity* (Cambridge: Cambridge University Press, 1989).

10. Ibid., 90.

11. Ernesto Laclau and Chantal Mouffe, *Hegemony and Socialist Strategy: Towards a Radical and Plural Democracy* (London: Verso, 1985), 183.

12. Ernesto Laclau, "Politics and the Limits of Modernity," in *Universal Abandon? The Politics of Postmodernism*, ed. Andrew Ross (Minneapolis: University of Minnesota Press, 1988), 66.

13. Ibid., 81.

14. Laclau and Mouffe, *Hegemony and Socialist Strategy*, 183.

15. Chantal Mouffe, "Democratic Citizenship and Political Community," in *Community at Loose Ends* (see note 9 above), 78.

16. Ibid., 79–82; Ernesto Laclau, "Universalism, Particularism, and the Question of Identity," *October* 61 (1992).

17. Laclau and Mouffe, *Hegemony and Socialist Strategy*, 184.

18. Ernesto Laclau, *New Reflections on the Revolution of Our Time* (London: Verso, 1990), 233.

19. Ibid., 79.

20. Laclau and Mouffe, *Hegemony and Socialist Strategy*, 191.

21. Ibid., 182.

22. In *Homegirls: A Black Feminist Anthology*, ed. Barbara Smith (New York: Kitchen Table/Women of Color Press, 1983).

23. Steven Lukes, "Of Gods and Demons: Habermas and Practical Reason," in *Habermas: Critical Debates*, ed. John Thompson and David Held (Cambridge: MIT Press, 1982).

24. Thomas McCarthy, *Ideals and Illusions: On Deconstruction and Reconstruction in Contemporary Political Theory* (Cambridge: MIT Press, 1991), 191.

25. For a compelling sense of possibility in the face of seemingly impossible situations, see Václav Havel, *Living in the Truth*, trans. Jan Vladislav (London: Faber and Faber, 1986).

26. Theodor Adorno, "Subject and Object," in *The Essential Frankfurt School Reader*, ed. A. Arato and E. Gebhardt (New York: Continuum, 1982), 499–500.

27. There is no more beautiful and powerful expression of Adorno's transfigurative relation to Hegel than that in "Aspects of Hegel's Philosophy," in Theodor Adorno, *Hegel: Three Studies*, trans. S. W. Nicholsen (Cambridge: MIT Press, 1993).

28. For a discussion of ways in which a tradition might cultivate a memory and an anticipation of profound engagements with other traditions, see Romand Coles, "Storied Others and Possibilities of Caritas: Milbank and Neo-Nietzschean Ethics," *Modern Theology* 8, no. 5 (1992), 331–51.

29. Though, and here things get too difficult for me to unravel, the reverse might be said as well.

30. Habermas reflects explicitly on this theme in his "Further Reflections on the Public Sphere," in *Habermas and the Public Sphere*, ed. Craig Calhoun (Cambridge: MIT Press, 1992).

31. Ibid., 444.

32. Ibid.

33. See John Dewey, *The Public and Its Problems* (Athens, Ohio: Swallow Press, 1954).

34. This argument is vital to Habermas's effort to counter arguments from "democratic realists" that the public is incapable of intelligently leading democracies—which therefore require elite guidance. Democratic realism in the twentieth century begins at least with Walter Lippmann's *Public Opinion* (1922) and *The Phantom Public* (1925). See Robert Westbrook, *John Dewey and American Democracy* (Ithaca: Cornell University Press, 1991), for an illuminating discussion of the debate between Dewey and Lippmann. As I read it, the debate has changed little in seventy years.

35. One of Habermas's clearest and most concise statements of this position, and of his argument for continuing the project of the welfare state with a "higher level of reflection," is to be found in "The New Obscurity: The Crisis of the Welfare State and the Exhaustion of Utopian Energies," in Jürgen Habermas, *The New Conservatism: Cultural Criticism and the Historians' Debate*, trans. S. W. Nicholsen (Cambridge: MIT Press, 1989).

36. I situate myself somewhere between Jean L. Cohen and Andrew Arato's overly optimistic account of the condition of civil society, in their comprehensive *Civil Society and Political Theory* (Cambridge: MIT Press, 1992), and Michael Hardt and Antonio Negri's overly pessimistic claim that "civil society has been reduced to an administrative mechanism," in *Labor of Dionysus: A Critique of the State Form* (Minneapolis: University of Minnesota Press, 1994), 300.

37. The following is an exploration of the possible implications of the ethic being developed here. It is not intended as a claim about the historical transformative potential of ethical philosophy. If it were, it would be highly exaggerated.

38. George Kateb, *The Inner Ocean* (Ithaca: Cornell University Press, 1992), develops some of these themes, though not in the directions I am suggesting.

39. Though both McCarthy ("Complexity and Democracy: The Seducements of

Systems Theory," in his *Ideals and Illusions*) and Cohen and Arato (chap. 9 in *Civil Society and Political Theory*) make compelling arguments against the rigid and dichotomous character of Habermas's employment of system and lifeworld. Perhaps *FN* begins to address this problem.

40. McCarthy, "Complexity and Democracy," 169.

41. Ibid., 180.

42. Ibid. Robert D. Putnam's *Making Democracy Work: Civic Traditions in Modern Italy* (Princeton: Princeton University Press, 1993) supports these claims. More theoretically, Bowles and Gintis, *Capitalism and Democracy*, and Benjamin Barber, *Strong Democracy* (Berkeley: University of California Press, 1984) make a similar case for the learning and coordinative capacities of participation. At the intersection of theory and public policy, John Dryzek, in *Discursive Democracy: Politics, Policy, and Political Science* (Cambridge: Cambridge University Press, 1990), elaborates this theme in terms of "discursive designs."

43. Putnam, *Making Democracy Work*, 117.

Index

Index 245

CONTESTATIONS

CORNELL STUDIES IN POLITICAL THEORY

A series edited by
WILLIAM E. CONNOLLY

Romand Coles is Associate Professor of Political Science at Duke University. He is the author of *Self/Power/Other: Political Theory and Dialogical Ethics*, also from Cornell.